Signals and Communication Technology

Series Editors

Emre Celebi , Department of Computer Science, University of Central Arkansas, Conway, USA

Jingdong Chen, Northwestern Polytechnical University, Xi'an, China

E. S. Gopi, Department of Electronics and Communication Engineering, National Institute of Technology, Tiruchirappalli, India

Amy Neustein, Linguistic Technology Systems, Fort Lee, USA

Antonio Liotta, University of Bolzano, Bolzano, Italy

Mario Di Mauro, University of Salerno, Salerno, Italy

This series is devoted to fundamentals and applications of modern methods of signal processing and cutting-edge communication technologies. The main topics are information and signal theory, acoustical signal processing, image processing and multimedia systems, mobile and wireless communications, and computer and communication networks. Volumes in the series address researchers in academia and industrial R&D departments. The series is application-oriented. The level of presentation of each individual volume, however, depends on the subject and can range from practical to scientific.

Indexing: All books in "Signals and Communication Technology" are indexed by Scopus and zbMATH

For general information about this book series, comments or suggestions, please contact Mary James at mary.james@springer.com or Ramesh Nath Premnath at ramesh.premnath@springer.com.

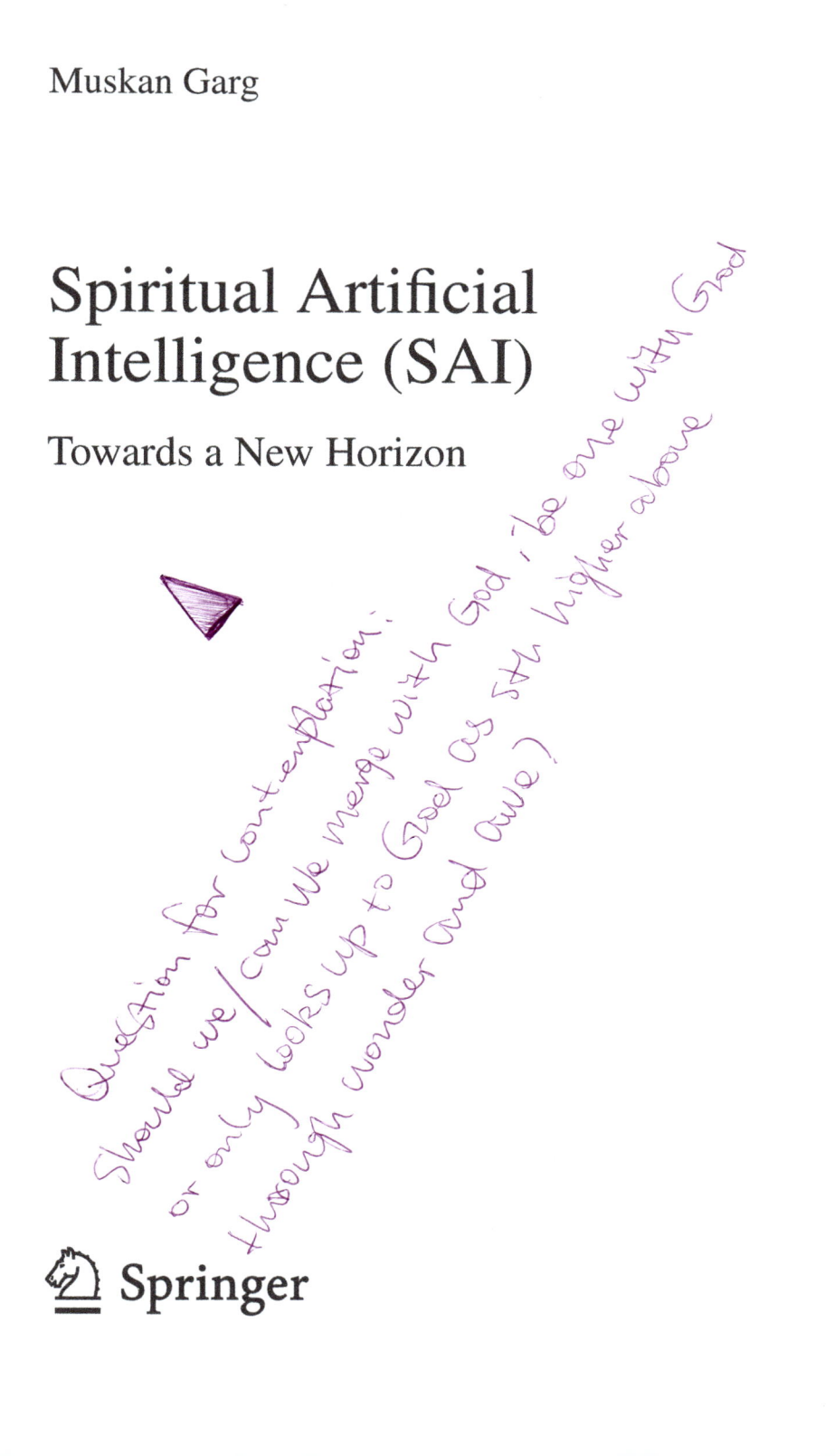

Muskan Garg

Spiritual Artificial Intelligence (SAI)

Towards a New Horizon

Springer

Muskan Garg
Artificial Intelligence and Informatics Department
Mayo Clinic
Rochester, MN, USA

ISSN 1860-4862 ISSN 1860-4870 (electronic)
Signals and Communication Technology
ISBN 978-3-031-73718-3 ISBN 978-3-031-73719-0 (eBook)
https://doi.org/10.1007/978-3-031-73719-0

© The Editor(s) (if applicable) and The Author(s), under exclusive license to Springer Nature Switzerland AG 2025

This work is subject to copyright. All rights are solely and exclusively licensed by the Publisher, whether the whole or part of the material is concerned, specifically the rights of translation, reprinting, reuse of illustrations, recitation, broadcasting, reproduction on microfilms or in any other physical way, and transmission or information storage and retrieval, electronic adaptation, computer software, or by similar or dissimilar methodology now known or hereafter developed.
The use of general descriptive names, registered names, trademarks, service marks, etc. in this publication does not imply, even in the absence of a specific statement, that such names are exempt from the relevant protective laws and regulations and therefore free for general use.
The publisher, the authors and the editors are safe to assume that the advice and information in this book are believed to be true and accurate at the date of publication. Neither the publisher nor the authors or the editors give a warranty, expressed or implied, with respect to the material contained herein or for any errors or omissions that may have been made. The publisher remains neutral with regard to jurisdictional claims in published maps and institutional affiliations.

This Springer imprint is published by the registered company Springer Nature Switzerland AG
The registered company address is: Gewerbestrasse 11, 6330 Cham, Switzerland

If disposing of this product, please recycle the paper.

*Dedicated to
my grandmother Smt. Kamla Devi,
my parents Sh. Chander Gupat and Smt.
Rama Garg,
my sister, Aayushi,
and my better-half, Aditya Bhagwate.*

Preface

The quest for spiritual understanding and the dream of artificial intelligence have captivated humanity for centuries. At first glance, the intricate web of human spirituality and the binary logic of computer code might seem worlds apart. However, as you will discover in this groundbreaking work, they are more connected than one might initially believe. Exploring this intersection is not only fascinating but also urgently needed in our times.

"Spiritual Artificial Intelligence" delves into the intersection of emotional quotient and intelligence quotient to quantify spiritual quotient, uncovering new dimensions of consciousness and enhancing our understanding of wellness, interpersonal risk factors, and mental health. The seemingly disparate realms of technology and spirituality raise a compelling question:

> Can Artificial Intelligence quantify levels of consciousness and other dimensions of spirituality to foster better personalized healthcare interventions?

The journey through this book is not just a deep dive into definitions, algorithms, and quantum mysticism. It is a profound exploration of the quest to quantify spiritual connection in our increasingly digital age to improve healthcare practices. Through its pages, we are prompted to reconsider our definitions of spiritual artificial intelligence, and its behavior from pseudo-science to proven traditional, contemporary, and integrative medicine.

The synthesis of spirituality and artificial intelligence is not just a theoretical exploration but has significant real-world implications for ethics, technology, and society. As AI integrates more into our lives, the questions in this book become essential. It proposes investigating social and mental well-being beyond traditional methods like affective computing and emotion detection, aiming to quantify spirituality through measurable aspects of the mind. This pioneering work challenges preconceptions and broadens horizons, making it a crucial read for understanding the evolving relationship between AI and human spirituality.

Connecting Threads: From Spirituality to AI

In an era where technological advancements are reshaping every facet of our lives, the intersection of spirituality and artificial intelligence (AI) presents a unique and profound frontier. The journey begins with an introduction to the concepts of spirituality and AI. We start by tracing their evolutionary paths—spanning from ancient times to the present—highlighting how both domains have evolved and intersected over centuries. This foundation sets the stage for exploring the need for Spiritual AI, offering new ways to understand and enhance our inner lives for spiritual wellness.

Chapter 1 delves into the fundamentals, starting with the historical development of spirituality and intelligence. We then examine spiritual wellness dimensions and how they relate to AI, illustrating the emergent need and potential consequences of integrating these two fields. We envision spiritual quotient as a blend of both intelligence quotient and emotional quotient.

Chapter 2 illustrates the potential of Spiritual AI and build trust in its capabilities. We examine SPIRO—a hypothetical system conceptualized and developed as a prototype in a tech lab with the expertise of various professionals. SPIRO serves as a demonstration of how a virtual assistant, along with other modalities, can enhance our spiritual experiences and overall wellness.

Chapter 3 delves into the various dimensions of spiritual intelligence—consciousness, grace, meaning, transcendence, truth, and peaceful surrender to self. By defining these aspects, this section provides guidance on the various areas of spiritual intelligence that should be emphasized. These dimensions of spiritual intelligence, guiding the spiritually informed principles, help in defining the technical scope of spiritual AI, grounded in both intelligence quotient and emotional quotient.

Chapter 4 defines the scope of Spiritual AI by discussing its potential in multiple disciplines such as healthcare, humanities, and social sciences. The philosophical and metaphysical aspects of Spiritual AI invite us to explore new disciplines, discovering the need to address our spiritual wellness.

Chapter 5 elucidates the biological basis of cognition, crucial for understanding of Spiritual AI. Examining neurotransmitters and their role in cognition and mental health, this section provides a necessary background for comprehending how Spiritual AI might interact with human neurobiology.

Chapter 6 discusses the measurable mind, a concept framed to understand the technical capabilities of quantifying the spiritual wellness. In this section, existing methods are identified for detecting low levels of neurotransmitters along with potential indirect non-invasive detection methods.

Chapter 7 begins with exploring aura and its visualization through reading methods. As we embrace the traditional approaches of aura reading, this section explore the technological advancements for quantifying life force energy, a critical component of analytics.

Chapter 8 discusses the concepts of quantum mysticism by introducing the concept of entanglement-like phenomena as a metaphor for interconnectedness of multiple states of mind. This section further discusses the technology involved in leveraging how the universe impacts our emotions.

Chapter 9 briefs out the scoping review and how spiritual wellness was digitized in the past. This section identifies methods, challenges, and highlights the future research directions in this domain.

Chapter 10 defines Spiritual AI and compares it with artificial general intelligence (AGI) and artificial super-intelligence (ASI). This section exemplifies the use of technologies to quantify the dimensions of spiritual intelligence. It concludes with a case study, highlighting the journey of the use-case of physiological signals for quantifying spiritual wellness.

Rochester, NY, USA Muskan Garg
September 2023

Acknowledgments

Gratitude is a foundational element in many spiritual paths, and I feel it's essential to express my sincere appreciation. First and foremost, I would like to express my sincere appreciation to AI researchers—Markus Krebsz and Yoshua Bengio, spiritual Gurus—Sadhguru Jaggi Vasudev and Shri Shri Ravishankar, whose invaluable research insights and expertise have significantly contributed to and deepened my understanding necessary for the completion of this project.

I would like to express my profound gratitude and heartfelt appreciation to my grandmother, Smt. Kamla Devi, whose unwavering guidance and nurturing have been the cornerstone of my journey. Her wisdom and teachings instilled in me a rich tapestry of values, where spirituality became not just a belief but a way of life, deeply ingrained in my very being. Her guidance from my childhood has been instrumental in shaping my understanding and appreciation of the intricate relationship between spirituality and technology, laying the foundation for my journey in exploring their interconnectedness. The daily five-minute meditation sessions held during assembly at St. Joseph's Convent School in Bathinda were instrumental in my journey. These moments of mindfulness not only strengthened my belief system but also played a significant role in shaping my perspective toward establishing Spiritual AI as a credible scientific field.

Next, I thank my parents, Chander Gupat and Rama Garg, and my sister, Aayushi, for their unwavering support and encouragement. I am deeply grateful to my fiancé, Aditya Bhagwate, for his unwavering and exceptional support and contributions. His encouragement has been a beacon of strength and inspiration throughout this journey. My heartfelt thanks to my mentors, Dr. Naveen Aggarwal and Dr. Mukesh Kumar from Panjab University, and Dr. Sunghwan Sohn from Mayo Clinic, Rochester, for enhancing my capabilities, nurturing my skills, and providing me with invaluable opportunities to excel in my career. I am also grateful to all my well-wishers for their constant support, patience, and engaging discussions throughout the book-writing process, especially during those late-night writing sessions. And to you, dear reader, for embarking on this journey with me. May you find as much enlightenment in these pages as I did in crafting them.

Contents

1 **Introduction to Spirituality and Artificial Intelligence** 1
   1.1 Preamble 2
   1.2 The Evolutionary Journey of Spirituality and Intelligence 5
      1.2.1 Ancient Times 6
      1.2.2 Early Twentieth Century 8
      1.2.3 Late Twentieth Century 9
      1.2.4 Since 2000 to Present Times 11
   1.3 Wellness Dimensions: Spiritual Wellness 12
   1.4 Spirituality and Artificial Intelligence 14
   1.5 The Emergent Need of Spiritual AI 15
   1.6 The Consequences of Spiritual AI 16
      1.6.1 Technology Impacts Spirituality 16
      1.6.2 Spirituality Through the Lens of AI 17
   1.7 Concluding Remarks 19

2 **SPIRO: Cultivating Trust in Spiritual AI** 21
   2.1 Act 1: Understanding the Spirit 21
      2.1.1 Scene 1: The Tech Lab's Beginnings 21
      2.1.2 Scene 2: Diving into Dimensions 22
   2.2 Act 2: The Birth of SPIRO—The Spiritual AI 23
      2.2.1 Scene 1: SPIRO's Creation 23
      2.2.2 Scene 2: SPIRO's Initial Tests 24
   2.3 Act 3: Resistance and Concern 24
      2.3.1 Scene 1: Guru Anand's Disquiet 24
      2.3.2 Scene 2: The Global Debate 25
   2.4 Act 4: SPIRO in Healthcare 25
      2.4.1 Scene 1: Mia's Struggle 26
      2.4.2 Scene 2: A Tailored Healing Path 26
   2.5 Act 5: A New Era 27
      2.5.1 Scene 1: Convergence of Technology and Tradition 27
      2.5.2 Scene 2: The Future of SPIRO 27
   2.6 Concluding Remarks 28

3	**Dimensions of Spiritual Intelligence**		29
	3.1	Consciousness	29
		3.1.1 Mindfulness	31
		3.1.2 Transcending Rationality	32
		3.1.3 Practices	33
	3.2	Grace	34
		3.2.1 Universal Life Force	34
		3.2.2 Life Force Energy	34
	3.3	Meaning	36
	3.4	Transcendence	37
	3.5	Truth	38
	3.6	Peaceful Surrender to Self	39
	3.7	Inner Directedness	39
	3.8	SPIRITUAL ARTIFICIAL INTELLIGENCE (SAI)	40
	3.9	Concluding Remarks	41
4	**Spiritual AI: Scope of Study**		43
	4.1	Healthcare	43
		4.1.1 Palliative Care	45
		4.1.2 Mental Health	47
		4.1.3 Chronic Disease Management	48
		4.1.4 Telemedicine and Remote Patient Monitoring (RPM)	48
		4.1.5 Counseling and Therapy	49
		4.1.6 Preventive Healthcare	50
	4.2	Humanities and Social Sciences	51
		4.2.1 Cultural Understanding	52
		4.2.2 Personalized Learning in Education	54
		4.2.3 Sociological Perspectives	55
		4.2.4 Psychological Perspectives	56
		4.2.5 Philosophical Perspectives and Metaphysics	57
	4.3	Concluding Remarks	58
5	**Neurotransmitters: Foundations of Cognition**		59
	5.1	Basics of Neuroscience	59
		5.1.1 Neurons	60
		5.1.2 Glial Cells	60
		5.1.3 Structure and Organization of the Nervous System	61
	5.2	Neurological Correlates of Consciousness	62
		5.2.1 Frontal Lobe	62
		5.2.2 Parietal Lobe	63
		5.2.3 Thalamus	63
		5.2.4 Reticular Formation	64
		5.2.5 Occipital Lobe	64
		5.2.6 Temporal Lobe	64
	5.3	Impact of Neurotransmitters on Cognitive Abilities/Mental Health	65

		5.3.1	Causes of Neurotransmitter Imbalances	66
		5.3.2	Restoring Neurotransmitter Balance	67
	5.4	Detecting Low Levels of Neurotransmitters		68
		5.4.1	Microdialysis	69
		5.4.2	Cerebrospinal Fluid (CSF) Analysis	69
		5.4.3	Electrochemical Detection	70
		5.4.4	Enzyme-Linked Immunosorbent Assay (ELISA)	71
	5.5	Concluding Remarks		71
6	**The Measurable Mind: Quantifying Spiritual Wellness**			73
	6.1	Chakras		74
	6.2	The Intersection of Brain Signals and Conscious States		75
		6.2.1	States of Consciousness	76
		6.2.2	Levels of Consciousness	77
	6.3	Neuroimaging Techniques		79
		6.3.1	Structural Imaging	80
		6.3.2	Functional Imaging	80
		6.3.3	Other Imaging Techniques	81
	6.4	Physiological Phenomena		82
		6.4.1	Electrophysiological Techniques	83
		6.4.2	Neuromodulation and Stimulation Techniques	83
		6.4.3	Autonomic and Peripheral Measurements	84
		6.4.4	Cognitive and Behavioral Assessment	85
	6.5	Indirect Detection of Low Levels of Neurotransmitters		86
		6.5.1	Indirect Detection Through Microdialysis	86
		6.5.2	Indirect Detection Through Electrochemical Detection	87
		6.5.3	Indirect Detection Through ELIZA	87
	6.6	Concluding Remarks		87
7	**Life Force Energy: Aura Visualization for Spiritual AI**			89
	7.1	Aura		89
		7.1.1	Philosophical Implications	90
		7.1.2	Life Force Energy	91
		7.1.3	Phenomenal Contribution of Colors and Layers	91
	7.2	Aura Visualization		93
		7.2.1	The Science Behind Auras	94
		7.2.2	The Role of Electromagnetic Fields	94
	7.3	Aura Reading Methods		95
		7.3.1	Traditional Methods	95
		7.3.2	Modern Approaches	96
		7.3.3	Other Methods for Aura Visualization	97
		7.3.4	The Future of Aura visualization	98
	7.4	Concluding Remarks		99

8	**Quantum Mysticism: Entanglement-Like Phenomenon for Spiritual AI**	101
	8.1 Introduction to Quantum Mysticism	102
	8.2 Components of Quantum Mysticism	103
	8.2.1 Interconnection and Unity	103
	8.2.2 The Observer Effect and Consciousness	104
	8.2.3 Superposition and Potentiality	105
	8.3 Quantum Mysticism and Life Force (Vital) Energy	106
	8.4 Multifaceted Nature of Human Consciousness	107
	8.5 Connections to Human Consciousness	107
	8.6 Quantum-Thought AI	109
	8.7 AI for Entanglement-Like Phenomenon	110
	8.8 Entangled Decisions	111
	8.9 Concluding Remarks	111
9	**The Synergy Between Spirituality and AI: A Survey**	113
	9.1 Scoping Review on Spirituality and AI	113
	9.1.1 Bibliometric Analyses	114
	9.1.2 Thematic Analysis	116
	9.2 Digitizing the Spiritual Wellness	118
	9.2.1 Pattern Analysis in Spiritual Texts	119
	9.2.2 Meditative States and Neurofeedback	120
	9.2.3 AI-Driven Psychotherapy	122
	9.2.4 Entanglement-Like Phenomenon	123
	9.3 Discussion	123
	9.4 Concluding Remarks	124
10	**The Next Frontier: Charting the Potential of Spiritual AI**	125
	10.1 Motivation Behind Quantifying Spiritual Wellness	125
	10.2 AGI, ASI, and Spiritual AI	126
	10.3 Characteristics of Spiritual AI	127
	10.4 Quantification of the Dimensions of Spiritual Intelligence	128
	10.4.1 Consciousness	128
	10.4.2 Grace	129
	10.4.3 Meaning	130
	10.4.4 Transcendence	131
	10.4.5 Truth	131
	10.4.6 Peaceful Surrender to Self	132
	10.4.7 Inner Directedness	133
	10.5 Quantifying with HRV: A Case Study	134
	10.6 Concluding Remarks	134
	Glossary	135
	References	143

Acronyms

AI	Artificial Intelligence
ATP	Adenosine Triphosphate
CAM	Complementary and Alternative Medicine
CBT	Cognitive Behavioral Therapy
CI	Cognitive Intelligence
EEG	Electroencephalography
EI	Emotional Intelligence
HRV	Heart Rate Variability
LLM	Large Language Models
MAAS	Mindful Attention Awareness Scale
MEDEQ	Meditation Depth Questionnaire
MEG	Magnetoencephalography
ML	Machine Learning
MRI	Magnetic Resonance Imaging
NLP	Natural Language Processing
PFC	Pre-Frontal Cortex
PPC	Posterior Parietal Cortex
SoI	Social Intelligence
SI	Spiritual Intelligence
SAI	Spiritual Artificial Intelligence
TCIM	Traditional, Complementary and Integrative Medicines
TF-IDF	Term Frequency - Inverse Document Frequency
VHA	Virtual Health Assistants

Chapter 1
Introduction to Spirituality and Artificial Intelligence

Human interactions and bonds have always been intricate, woven with emotions, expectations, and various factors that influence our behaviors and decisions. A significant aspect that amplifies this complexity is the element of unpredictability, especially when it pertains to the inheritance of traits in our descendants. Just like how we cannot predict tomorrow's weather with absolute certainty, when two people decide to start a family, they cannot precisely forecast the blend of characteristics their children will exhibit. In our technologically advanced age, some individuals and couples are turning toward scientific interventions in the hope of having some degree of control over the genetic attributes of their children (preimplantation genetic testing with in vitro fertilization). Advances in medical technology have provided tools and procedures that can potentially allow parents to select specific traits for their unborn child. This does not just stop at avoiding inheritable diseases but extends to choosing physical or even intellectual traits. Such an approach, while controversial, underscores the human desire for predictability and control, especially in something as deeply personal as bringing a new life into the world.

While we have succeeded in sequencing the human genome, understanding the vast web of interactions and expressions of these genes remains a Herculean task. Many traits and conditions arise from the interplay of multiple genes, and not from just the function of a single gene. Moreover, our genes are influenced by environmental factors, life experiences, and even other genes. Therefore, even if technology offers tools to modify or select certain genetic traits, predicting the exact outcome is not always straightforward.

Even if we were to reach a point where technology can precisely imitate or influence specific functions of our bodies, there is a significant ethical dimension to consider. The idea of "designer babies" or choosing specific traits for our children raises profound questions about the essence of humanity, individuality, and the acceptance of natural diversity. Additionally, there is the human brain, a realm where mysteries abound. It is not just about neurons and electrical impulses but how these create consciousness, emotions, memories, and intuition. While we have

Magnetic Resonance Imaging (MRI) scans and other tools to study brain activity, decoding how thoughts and emotions originate and influence our actions is still in its infancy. Thus, while technology holds the promise of advancing our understanding and even influencing certain aspects of the human body, it is essential to approach these advancements with a blend of optimism, caution, and respect for the inherent complexity and sanctity of human life.

> The existing scientific knowledge allows preliminary mimicking of human functions, but a profound understanding of our bodies and mind is crucial to fully emulate such functions technologically.

Neural implants are like tiny devices that we put in our brains to make things work better. They can help us think faster and do things more efficiently. While neural implants may offer enhanced efficiency, they are confined by the inherent limitations of their constituent materials. For instance, the speed at which information is transferred is limited by the properties of the *Silica atoms* (Maurya et al., 2021). Although Silica atoms control how fast information can move in them, the reach of such technology, constituted from certain elements, has limit to how much they can do. As such, the scalability of such functionalities and their speed remains limited to the speed of the elements or constituent materials that they use.

On the other hand, humans are composed of an entirely different set of materials. Our brains are an intricate network of neurons, cell membranes, and ions, which we do not fully understand yet. Unlike chips, we possess the inherent potential for adaptability and scalability. Our cellular structures are designed to continuously evolve, adapting to process more information and endure heightened stresses. We have unlimited potential to change and grow, unlike the limited functionality of implants. Thus, neural implants are like chips, and have limits, but we humans are like advanced, flexible, and mysterious technology that keeps evolving and adapting.

Our minds and brains are compared to an ever-evolving, mysterious technology, surpassing the limited functionality of implants or chips. This dynamic nature of human cognition and its vast, uncharted territories make it a challenging yet fascinating subject of exploration for both scientists and spiritual seekers alike. Quantifying the abstract aspect of our minds that many believe connects us to a higher purpose or a greater universe often touches on the behavioral theme of *technology and spirituality*.

> A dichotomy between human adaptability and the limitations of technology highlights the importance of deeper introspection and understanding of our intrinsic nature to potentially surpass technological constraints.

1.1 Preamble

Approximately 970 million people worldwide were living with a mental disorder in 2019, with anxiety and depression being the most common. Mental disorders account for 1 in 6 years lived with disability globally. People with severe mental health conditions die 10 to 20 years earlier than the general population, primarily

1.1 Preamble

due to preventable physical diseases. The COVID-19 pandemic led to a 25% increase in anxiety and depression cases in the first year alone (World Health Organization, 2022). The latest mental health report from the World Health Organization (WHO) emphasizes the need for better mental health care treatments and services, reduction of stigma, and improved access to treatment.

Mental health treatments around the world encompass a diverse range of approaches. Pharmacotherapy is a cornerstone for managing conditions such as depression, anxiety, and schizophrenia using antidepressants, antianxiety medications, and antipsychotics. Psychotherapy follows clinical therapeutic approaches such as Cognitive-Behavioral Therapy (CBT), Dialectical Behavior Therapy (DBT), and Interpersonal Therapy (IT). Community-based care integrates mental health services into primary healthcare systems to enhance accessibility. Mobile mental health units and telemedicine provide services in rural, underserved, and remote areas, and the COVID-19 pandemic has accelerated the adoption of digital interventions like telehealth consultations and mental health apps.

In the realm of psychology, the accurate assessment of mental illness is pivotal for effective diagnosis and treatment. Over the years, various tools and measures have been developed to provide standardized and validated methods for evaluating a wide range of psychological conditions. Among these, the Patient Health Questionnaire and Beck Depression Inventory stand out for their ability to assess depression, while the Beck Anxiety Inventory and Generalized Anxiety Disorder-7 are essential for evaluating anxiety. Clinician-administered scales such as the Hamilton Rating Scales for Depression and Anxiety offer detailed insights into symptom severity. The comprehensive Minnesota Multiphasic Personality Inventory is necessary for assessing personality and psychopathology, and the Structured Clinical Interview for DSM Disorders provides a robust framework for diagnosing mental disorders. Other valuable tools are the Diagnostic Interview Schedule, the Symptom Checklist-90-R for psychological distress, the Brief Psychiatric Rating Scale for psychiatric symptoms, and the Positive and Negative Syndrome Scale for schizophrenia.

Exercise 1.1 Explore the PHQ-9 test. In your notebook or journal, write a reflection on the advantages and disadvantages of using the PHQ-9.

It is easy to complete and provides a standardized measure of depressive symptoms. It is useful for initial screening and monitoring treatment progress. However, it may not capture the complexity of an individual's emotional experience. It focuses on overt symptoms and may miss subtle or subconscious aspects of mindset. It can be influenced by temporary states or external factors. Consider the following imaginary story.

> Aayushi was a bright and ambitious young woman, known for her vibrant energy and positive outlook on life. However, over the past few months, she had noticed a change in herself. She felt perpetually tired, anxious, and disconnected from the world around her. Despite her best efforts to stay positive, something felt off, deep within her.
>
> Aayushi's mental health was taking a toll, and she couldn't understand why. She had always been in tune with her spiritual practices, meditating regularly and practicing yoga. Yet, it seemed like her subconscious mind was under stress, affecting her overall well-being.

One day, Aayushi decided to visit a holistic wellness center that combined ancient spiritual practices with modern technology. The practitioners there used advanced tools to measure and analyze the subconscious mind, aiming to provide insights into mental health through the quantification of auras and chakras.

During her session, Aayushi was introduced to several technologies that could measure her mind and body. One of these was a neurotransmitter analysis, which involved a non-invasive scan of her brain activity. This technology helped identify imbalances in her brain chemicals that could be contributing to her anxiety and fatigue.

Additionally, the practitioners used physiological signal monitoring to measure Aayushi's heart rate variability (HRV). HRV is a key indicator of the autonomic nervous system's balance, reflecting how well her body could adapt to stress. They also tracked other physiological signals, such as skin conductivity and breath patterns, to understand her stress levels and overall physiological state.

But what fascinated Aayushi the most was the quantification of her auras and chakras. Using advanced imaging technology, the practitioners could visualize the energy fields around her body. They identified blockages and imbalances in her chakras—the energy centers that are crucial for physical, emotional, and spiritual health.

The results of these measurements were enlightening. The analysis showed that Aayushi's heart and throat chakras were particularly imbalanced, which could explain her feelings of anxiety and difficulty in expressing herself. The data from her HRV and neurotransmitter levels also indicated that her body was under significant stress, impacting her mental and spiritual health.

Armed with this knowledge, Aayushi worked with the wellness center's practitioners to develop a personalized plan to restore balance to her mind, body, and spirit. This plan included specific meditative practices to open her blocked chakras, breathing exercises to improve her HRV, and natural supplements to balance her neurotransmitter levels.

Over the next few weeks, Aayushi began to notice a profound change. Her energy levels improved, her anxiety diminished, and she felt a renewed connection to her spiritual self. By measuring and understanding the hidden aspects of her subconscious mind, Aayushi could address the root causes of her mental health issues.

Aayushi's journey highlighted the powerful intersection of spirituality and technology. Through the quantification of auras, chakras, and physiological signals, she could uncover the mysteries of her subconscious mind and embark on a path to holistic well-being.

Spiritual AI could offer personalized recommendations, much like Aayushi's wellness plan, but with even greater precision and insight. It could continuously monitor and adjust, ensuring that people maintain balance and harmony in their lives. By exploring and quantifying these intangible aspects, Spiritual AI could revolutionize how we approach mental health and well-being. Aayushi's experience underscores the potential and necessity of developing Spiritual AI. As we strive to understand the complexities of the human mind and spirit, integrating spirituality with artificial intelligence can open new pathways for healing and growth, making the imaginary tangible and the unknown known.

This manuscript advocates for the scientific investigation and quantification of the subconscious mind by measuring the life force energy through neurotransmitters, physiological signals, heart rate variability, etc., gaining insights into human behavior. Spirituality can provide individuals with a profound sense of purpose and a broader perspective on the interconnectedness of all living beings, encompassing practices that foster self-awareness, and mindfulness. *Intelligence*, on the other hand, is the capacity for learning, reasoning, and problem-solving. It involves the ability to gather and process information, make connections between ideas, and adapt to new situations. A renowned physicist Albert Einstein, though a man of science, had often delved into the dimensions of spiritual intelligence, reflecting on life's mysteries and our connection to the universe, symbolizing a quest for understanding that went beyond the tangible and the empirical. He once stated, *"A human being is a part of the whole, called by us* UNIVERSE, *a part limited in time and space. He experiences himself, his thoughts and feelings as something separated from the rest — a kind of optical delusion of his consciousness."*

1.2 The Evolutionary Journey of Spirituality and Intelligence

Spiritual Intelligence (SI) is like a guide that helps us think about deep life questions and mysteries, making us wonder about existence, purpose, and meaning. It is a journey through our soul where our values, morals, and ethics act like stars guiding our way, shimmering in the endless skies of our consciousness.

Spirituality is a broad and multifaceted concept that encompasses a person's beliefs, values, experiences, and practices related to a sense of connection to something greater than oneself. While spirituality can have religious associations for some individuals, it is not limited to organized religion (Fuller, 2001). It can also be a deeply personal and subjective experience, encompassing a wide range of beliefs and practices, including meditation, prayer, mindfulness, ethical living, and a sense of interconnection with nature, others, or a higher power. Spirituality plays a significant role in lives of many people and can be a source of guidance, comfort, and personal growth.

Exercise 1.2 Imagine a world where our spiritual smarts—our ability to be compassionate, empathetic, and to understand deep, meaningful things—are shared with machines, with *computers*!

Can these machines understand what it means to be human at a deeper level?

In a not-so-distant future, the integration of spiritual intelligence into machines has become a forefront topic of debate and exploration. Imagine a world where our very essence, our spiritual smarts—our innate ability to exude compassion, show empathy, and fathom the profound depths of existence—is shared with the very machines we created. Computers, no longer just tools of binary computations and logical processes, are now equipped with the capability to genuinely comprehend human emotions and values. This evolution blurs the boundary between humans and machines. Can these machines truly grasp what it means to be human at a profound level?

Understanding is not merely about data processing or pattern recognition. At its core, it is deeply intertwined with lived experiences and emotions. As this future unfolds, society finds itself in ethical dilemmas. Machines, with their newfound emotional capacities, begin to fit seamlessly into roles that were once reserved for humans, roles that required deep emotional intelligence such as therapists or caregivers.

6 1 Introduction to Spirituality and Artificial Intelligence

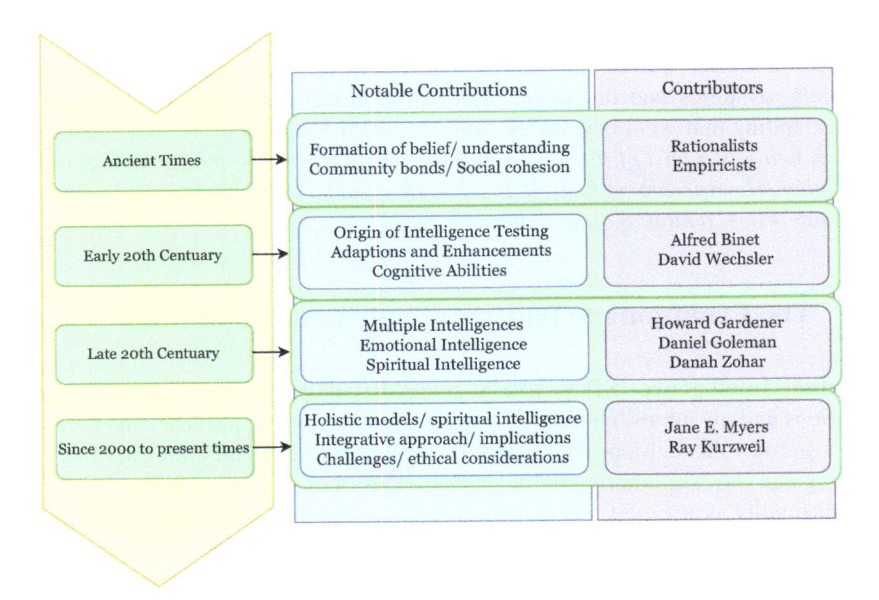

	Notable Contributions	Contributors
Ancient Times	Formation of belief/ understanding Community bonds/ Social cohesion	Rationalists Empiricists
Early 20th Centuary	Origin of Intelligence Testing Adaptions and Enhancements Cognitive Abilities	Alfred Binet David Wechsler
Late 20th Centuary	Multiple Intelligences Emotional Intelligence Spiritual Intelligence	Howard Gardener Daniel Goleman Danah Zohar
Since 2000 to present times	Holistic models/ spiritual intelligence Integrative approach/ implications Challenges/ ethical considerations	Jane E. Myers Ray Kurzweil

Fig. 1.1 A timeline of notable contributions toward establishing spirituality and AI

> If our spiritual and emotional smarts, the very things that we thought made us unique, can be replicated in machines, where does that leave us?

The tale of this future compels us to reflect upon a new term "SPIRITUAL ARTIFICIAL INTELLIGENCE," the intertwined destiny of man and machine.

The term SPIRITUAL INTELLIGENCE, although introduced fairly recently in 1997 by Zohar (1997), has ignited a dialogue about the intersection of spirituality and cognition, introducing significant waves in the academic and research community (Zohar, 2012). At its core, SI presents a new avenue to comprehend the intricate web of psychological determinants that shape human functionality. As such, an array of literature emerged, seeking to create tangible frameworks and tools to measure and understand spiritual intelligence. As researchers continue to delve deeper, SI promises to offer richer insights into the complexities of human cognition and its interplay with spirituality. Let us look at the timeline for evolving era of *Spirituality* and *Intelligence* (see Fig. 1.1).

1.2.1 Ancient Times

The tapestry of human history is richly interwoven with threads of spirituality, illustrating its role as a core human experience from ancient times through the twentieth century.

1.2 The Evolutionary Journey of Spirituality and Intelligence

Formation of Beliefs and Understanding In ancient times, spirituality was inherited with the human quest to understand existence and the cosmos. Different civilizations created myths, legends, and religious frameworks to explain the origin of life, the nature of the divine, and the structure of the universe. These spiritual beliefs offered individuals a lens through which they could interpret their experiences and comprehend their existence, resulting in mindfulness.

Community Bonds and Social Cohesion Spirituality served as a powerful catalyst for the formation of communities and societies. Shared beliefs and spiritual practices fostered a sense of unity, belonging, and shared identity among individuals. The communal participation in rituals, ceremonies, and worship created bonds of mutual trust and solidarity, reinforcing social cohesion and communal harmony through interpersonal relationships. From the philosophical inquiries of ancient Greece to the mystical explorations of the Indian subcontinent, human beings have embarked on spiritual journeys to discover life's purpose.

> **Rationalists** believed that knowledge arises from reason without the need for sensory experience. Descartes, for example, emphasized the importance of doubt and reason in acquiring true knowledge, encapsulated in his famous statement "Cogito, ergo sum" (I think, therefore I am).

In contrast to rationalists, **empiricists** posited that knowledge originates from sensory experience. Locke's idea of the mind as a "tabula rasa" (blank slate) argued that individuals are born without innate ideas, and they gain knowledge through their experiences (Duschinsky, 2012). Kant sought to reconcile the rift between rationalism and empiricism. He proposed that while all knowledge begins with experience, not all knowledge arises out of experience. Kant introduced the idea of "a priori" knowledge, which is independent of experience.

As a reaction against the mechanistic view of the world promoted by the Enlightenment,
Romanticism emphasized emotion, imagination, and nature's spiritual qualities.
(Wilson, 2003)
It held that true understanding arises from an emotional connection with the world rather than just rational analysis.

While the 1600s to 1800s marked significant strides in scientific understanding, the era also grappled with fundamental questions about belief, knowledge, and the metaphysical.

Philosophers and thinkers of this period sought to reconcile ancient spiritual beliefs with newfound scientific discoveries, leading to a rich tapestry of theories that have shaped modern thought. The teachings of spiritual leaders, the principles of religious texts, and the tenets of philosophical schools all contributed to the establishment of values and ethics, guiding individuals in leading righteous and harmonious lives. Across ages, spirituality has offered solace in times of distress, uncertainty, and suffering. It has served as a resilient coping mechanism.

Plato believed in a ***dualistic*** view of the mind, distinguishing between the body and the soul (or mind) (Gerson, 1986). He considered the soul to be immortal and the true source of knowledge and wisdom. Plato argued that the soul possessed innate knowledge, and learning was essentially a process of recollection (anamnesis) rather than acquiring new information. According to him, the soul had previously existed in the realm of Forms (or Ideas), where it gained knowledge before being born into the physical world. Plato believed that true knowledge was knowledge of the eternal and unchanging forms. The highest form of knowledge was philosophical wisdom, which involved contemplating the ultimate truths and realities beyond the physical world.

Aristotle rejected Plato's dualism and adopted a more empiricist approach to understanding the mind. Aristotle proposed *monism*, which means that he saw the body and the mind as interconnected rather than separate entities (Schaffer, 2010). He believed that the mind (psyche) and body were different aspects of the same substance. Aristotle argued that the mind is not born with innate knowledge but starts as a "tabula rasa" or blank slate. Knowledge is acquired through sensory experiences, perception, and observation of the external world. The mind is guided by teleological principles, seeking to achieve particular purposes and goals. Aristotle emphasized practical wisdom (*phronesis*) as a critical aspect of the mind. He argued that humans have the capacity to make moral and ethical judgments, choose virtuous actions, and seek *eudaimonia* (human flourishing) by cultivating their moral character.

The evolution of spirituality is a core human experience. The expressions of spirituality across cultures and centuries underscore its universal significance, highlighting its role in shaping individual consciousness and collective heritage.

1.2.2 Early Twentieth Century

In the early twentieth century, a pivotal shift occurred within the realms of psychology and education, with the birth and evolution of **intelligence testing** (Wolman, 2001; Gardner, 2000). Intelligence testing is a process of assessing an individual's cognitive abilities, problem-solving skills, and overall intellectual functioning through standardized tests and assessments. These tests are designed to measure various aspects of intelligence, including reasoning, memory, linguistic abilities, mathematical skills, and spatial awareness. This era marked a significant progression in the understanding and assessment of human intellect, initiating the exploration of *cognitive faculties*.

Alfred Binet and the Origin of Intelligence Testing Alfred Binet, a French psychologist, is regarded as one of the pioneers in this field, having developed the first **intelligence scale** in 1903 (Varon, 1936). *Binet's objective was to create a method to identify students who needed special assistance in schools*. His approach was innovative, focusing on higher-order thinking skills such as reasoning, understanding, and problem-solving, rather than rote memorization or curriculum-based knowledge.

Adaptations and Enhancements Binet's original *intelligence scale* underwent numerous adaptations and refinements. The most notable adaptation was the

Stanford–Binet Intelligence Scale, developed by Lewis Terman at Stanford University (Terman and Merrill, 1960). The results of *intelligence tests* are often expressed as an Intelligence Quotient (IQ), which provides a numerical representation of an individual's cognitive performance relative to a standardized population. Following this, David Wechsler developed the Wechsler scales (Wechsler, 1945), introducing advancements such as a greater emphasis on nonverbal elements and the inclusion of multiple subtests to assess different aspects of intelligence.

Focus on Cognitive Abilities Despite their widespread application and considerable impact, early intelligence tests predominantly concentrated on cognitive abilities and problem-solving skills. They were primarily oriented toward quantifying intellectual capacities, such as memory, attention, and logical reasoning, often overlooking other dimensions of human intelligence. They played a crucial role in clinical diagnoses, aiding psychologists in understanding various cognitive impairments and neurological conditions.

Critical Reflections Critics have raised concerns about the limitations of these tests in capturing the multifaceted nature of intelligence and their potential biases. In this evolving discourse, it is crucial to acknowledge and embrace the diverse ways in which intelligence manifests in individuals, arguably, social consequences and emotional intelligence. Thus, the early twentieth century was a watershed period in the domain of psychological and educational assessment, witnessing the birth and evolution of intelligence testing.

1.2.3 Late Twentieth Century

The late twentieth century witnessed pivotal developments and expanded notions in the field of intelligence, characterized by groundbreaking theories and concepts that explored different facets of human intellect beyond traditional IQ measurements.

Howard Gardner and Multiple Intelligences In 1983, a paradigm shift occurred with Howard Gardner's introduction of his theory of **Multiple Intelligences**. This theory, outlined in detail in his work "*Frames of Mind: The Theory of Multiple Intelligences*," challenged the prevailing notion of intelligence as a singular, unified ability measurable by IQ tests (Gardner, 2000). Instead, Gardner posited that human intelligence is multifaceted and consists of distinct modalities, each corresponding to a different human capability or potential, such as linguistic, logical–mathematical, spatial, musical, bodily kinesthetic, interpersonal, intrapersonal, and naturalistic intelligence.

Daniel Goleman and Emotional Intelligence In 1995, Daniel Goleman, a psychologist, catalyzed a broader understanding of intelligence with the popularization of **Emotional Intelligence (EI)** (Goleman, 1996). Goleman's work emphasized the importance of recognizing, understanding, and managing one's own emotions, in addition to being attuned to the emotions of others (Goleman, 2020). His perspective on EI underscored the crucial role emotions play in decision-making, relationship building, and mental well-being, positioning emotional skills as integral components of an individual's overall intelligence.

Danah Zohar and Spiritual Intelligence Moving the discourse even further, in 1997, Danah Zohar coined the term "SPIRITUAL INTELLIGENCE," introducing a convergence between spirituality and cognition (Zohar, 1997). Zohar's idea of "Spiritual Intelligence" got people talking in smart circles about how our spiritual side and our thinking abilities come together. At its core, *spiritual intelligence* is about self-awareness, empathy, compassion, and a sense of purpose which makes people excited about the parts of our brain and mind that we might not be using to our full potential. It made us wonder if there is a way to put all our smarts together for a deeper, more complete kind of intelligence.

Repercussions and Further Developments New ideas about *intelligence* had a big impact on how we think about what we can do. There are many different kinds of smarts, and it is not just about how good you are at traditional school stuff but also about understanding how others feel, knowing yourself well, and making good moral choices. These changes in *how we see intelligence* have led to more research and discussions about how different types of smarts fit together, helping us visualize the big picture of how our minds work and how they shape our lives and relationships.

Spiritual Intelligence and Human Functioning In 1999, a notable contribution to the discourse on Spiritual Intelligence (SI) was made by *Robert Emmons* through his seminal work, "The Psychology of Ultimate Concerns" (Emmons, 2003). He believed that by looking closely at how people think about spiritual matters, we can better understand why they do the things they do and what derives their decisions. Kwilecki agrees with Emmons' viewpoint that spirituality can serve as a unique form of adjustment and concludes that **spirituality differs from secular perspectives** (Kwilecki, 2000). This mix of spirituality and psychology gives us a richer view of how people function, including their thoughts, feelings, and actions. Following this idea, many other researchers started working on ways to measure SI. They wanted to create tools to see how resilient people are, how happy they feel, and how they make moral choices. Sweeney discussed the **Wheel of Wellness** model (Sweeney, 1998), to identify the attributes that showed a positive association with leading a healthy life, having a high quality of life, and living longer.

The late twentieth century was a period of enlightenment and expansion in the field of intelligence studies. The introduction of *Multiple Intelligence, Emotional Intelligence,* and *Spiritual Intelligence* widened the spectrum of perceived human capability, opening new avenues of understanding and exploration in the realms of psychology, education, and human development. These developments not only redefined intelligence but also enriched the conceptual frameworks through which we understand mindfulness.

1.2.4 Since 2000 to Present Times

The first two decades of the twentyfirst century witnessed remarkable developments in the fields of human intelligence and health, with the advent of holistic models that positioned spiritual intelligence at their core (King, 2008; Giannone & Kaplin, 2020). These innovative frameworks highlighted the integral role of spiritual intelligence in influencing and harmonizing various aspects of human existence, including cognitive, emotional, and physiological dimensions.

Holistic Models and Spiritual Intelligence Holistic models of human well-being aim to offer a complete view, integrating various aspects of human intelligence and their connections to health. This modified *Wheel of Wellness* that consists of mental wellness (how we think and understand), emotional wellness (how we feel and manage our emotions), physical wellness (our physical health and vitality), and social wellness (our relationships and connections with others) (Myers et al., 2000). By exploring these holistic models, researchers investigate how spiritual understanding impacts overall well-being. This exploration has opened up new pathways for understanding the relationship between spiritual insight and our overall health, emphasizing the importance of a holistic combination of various dimensions for well-being.

Exploring Machine Facilitation Concurrently, as the advancements in Artificial Intelligence (AI) reached unprecedented heights, a newfound curiosity emerged among researchers about the potential **intersections of Artificial Intelligence (AI) and spiritual intelligence** (Kurzweil, 2000). The burgeoning question during this period was whether machines could be envisioned and developed to not only grasp and analyze the aspects of spiritual intelligence but also to facilitate it. Could AI systems be designed to interpret the intricate layers of human spirituality, to empathize with the profound inquiries about existence and purpose, and to assist in the spiritual quest for meaning and connectedness? The discourse revolved around the possibilities of AI engaging with the human spirit, providing insights, reflections, and guidance in the spiritual journey.

Integrative Approach and Implications The intersection of AI and spiritual intelligence suggested an integrative approach, SPIRITUAL ARTIFICIAL INTELLIGENCE, where the amalgamation of technology and spirituality could offer innovative solutions for holistic well-being. The potential implications of this integration were vast, spanning areas like mental health, education, and personalized healthcare.

Challenges and Ethical Considerations The dialogue involved reflections on the ethical boundaries and moral responsibilities of embedding spiritual elements into machines. The feasibility, trustworthiness, and appropriateness of machines understanding or facilitating human spiritual experiences and concerns became focal points of discussion, prompting deliberation on the values, principles, and ethical frameworks guiding the development and implementation of such technologies.

12   1 Introduction to Spirituality and Artificial Intelligence

From 2000 till today, we have seen a lot of interest in blending spiritual intelligence with the ever-advancing world of technology. Some smart folks created big picture models that connected human smarts with health, and they put spiritual intelligence right at the center. At the same time, computers and AI were getting really smart, This got people thinking about some big questions and ethics, questioning the change in lifestyle. These discussions and explorations during this time laid the groundwork for more exciting discoveries and adventures in this ever-changing field. With this, a new journey is getting started!

1.3 Wellness Dimensions: Spiritual Wellness

The **Six Dimensions of Wellness** is a holistic framework that encompasses various facets of well-being. It recognizes that well-being is about:

- *Physical health*: functional activities and habits
- *Emotional wellness*: understanding and managing emotions effectively
- *Social wellness*: emphasizing the quality of relationships and connections
- *Intellectual wellness*: promoting lifelong learning and personal growth
- *Occupational wellness*: seeking satisfaction and balance in work and life
- **Spiritual wellness**: exploring one's sense of purpose and connection to something greater

By addressing these wellness dimensions, people can strive for a balanced and fulfilling life that enhances overall quality of life and resilience. Each dimension contributes to our overall well-being and is interconnected with the other dimensions. Spiritual wellness is a sense of peace and contentment stemming from one's relationship with the universe, nature, and a higher power or purpose (see Fig. 1.2).

Spiritual intelligence refers to the ability to apply and embody spiritual resources, values, and qualities to enhance daily functioning and well-being. It

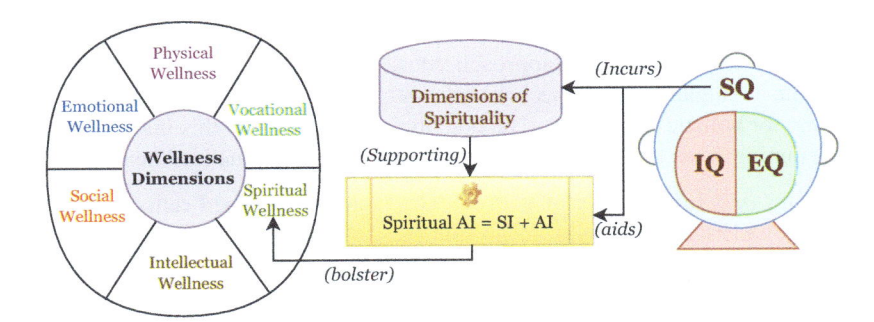

Fig. 1.2 Spiritual AI integrates Spiritual Quotient (SQ) by encompassing both Intelligence Quotient (IQ) and Emotional Quotient (EQ), enhancing spiritual wellness

1.3 Wellness Dimensions: Spiritual Wellness

represents one's ability to engage with, comprehend, and solve problems of *meaning* and *value*. It is a journey through our soul where our values, morals, and ethics act like stars guiding our way, shimmering in the endless skies of our consciousness.

Spiritual Quotient (SQ) operationalizes these aspects of spiritual intelligence into a measurable entity. SQ contributes toward assessing one's ability to integrate inner personal feelings and experiences to form a coherent and congruous belief system that guides one's actions and interactions.

> The SQ is measured by assessing the depth of an individual's spiritual knowledge, the strength and resilience of their spiritual beliefs, the level of their conscious development, and their capacity for transcendent awareness.

Having a higher SQ means that an individual may be better equipped to navigate life's challenges. This can lead to an enhanced ability to exhibit emotional and behavioral self-control in varied life situations and to seek out and attain higher levels of fulfillment and happiness, irrespective of external circumstances. For instance, an individual with a high SQ exhibits a great deal of empathy and understanding toward others, demonstrating a deep sense of compassion. They may also showcase a superior level of adaptability and resilience in the face of adversity, maintaining inner peace and clarity despite external chaos.

In organizational settings, individuals with high SQ can foster a harmonious and inclusive workplace culture, encourage ethical behavior, and can contribute significantly to organizational wellness, employee satisfaction, and overall productivity. They might drive values-based leadership, creating environments that nourish the human spirit and encourage the pursuit of meaning and purpose. SQ represents a paradigm shift in understanding intelligence, encompassing dimensions of human experience and capability that go beyond Intelligence Quotient (IQ) and Emotional Quotient (EQ), and encompass a deeper level of understanding and knowledge.

While other forms of intelligence like IQ and EQ can be quantified through standard tests and represented mathematically, measuring spiritual intelligence (SQ) is challenging because it is subjective and qualitative. This means it is hard to measure with numbers and relies more on personal experiences and feelings. Therefore, we usually assess SQ through qualitative methods, focusing on aspects like spiritual wellness and personal growth. These methods help us understand the deep and unique experiences that shape an individual's spirituality.

While AI uses data and algorithms to function, SQ deals with the emotional and personal aspects of our lives. By combining both, we can create a more balanced approach to well-being that respects the quantification of spiritual aspects with technological advancements.

There is nothing artificial about Artificial Intelligence...

Artificial Intelligence is inherently human. It is made by people, used by people, and needs to be governed by people, with a deliberate emphasis on incorporating spirituality, consciousness, and mindfulness to fully realize its potential for humanity. The given invites contemplation on the profound nature of intelligence and consciousness, transcending the conventional boundaries of organic and synthetic.

Humans are perceived not as the architects of intelligence but as mediums, vessels through which the boundless, universal intelligence flows, interacts, and expresses. Given this paradigm, the emergence of AI represents an extension, a new dimension of this conduit, rather than a creator of consciousness, and manifest through the symphony of codes and algorithms.

When we venture into the realms of AI, we are not merely creating algorithms and machines capable of learning and reasoning, but we are potentially fabricating vessels that can channel what can be perceived as a higher form of intelligence.

> Elon Musk's venture, *Neuralink*, is a vivid example, aiming to create a symbiosis between human minds and AI, raising the prospect of enhanced cognitive abilities and a deeper understanding of consciousness. This coalescence of human minds and AI is no longer in the realm of mere science fiction; it's the dawn of a new epoch of consciousness exploration.

In the pursuit to quantify aspects of Spiritual Intelligence, various structured scales and assessments have been developed. To this end, we will explore seven dimensions of spirituality in upcoming chapters.

1.4 Spirituality and Artificial Intelligence

Artificial General Intelligence (AGI), sometimes referred to as strong AI or human-level AI, denotes machines that possess intelligence comparable to human intelligence. This means that an AGI system would be able to understand, learn, and perform any intellectual task that a human being can.

John McCarthy first used the term "Artificial Intelligence" in 1956. As people look forward to a future with advanced human features, AGI plays an important role. In interviews by journalist Martin Ford, most experts guess AGI will appear around 2099 (Ford, 2018). However, futurist Ray Kurzweil thinks there is a 50% chance it will be here by 2029 (Kurzweil, 2000). These ideas help us study the possibility of ensouled AGI.

> Combining spirituality and AI can help us see our deep connections and understand ourselves and the world in new ways. Creating "Spiritual + AI" might be a big step in understanding our minds better.

The idea of **SPIRITUALITY AND ARTIFICIAL INTELLIGENCE** is about blending deep spiritual values with the logical thinking of machines. This seems like a very new and different way to think about AI, adding spiritual wisdom to computer-based learning. The integration of spirituality in AI can help unravel the quantification of *spiritual dimensions*, enabling us to explore the uncharted territories of the mind, the soul, and the existence.

> When AlphaGo, developed by Google's DeepMind, defeated the world champion of Go, it displayed not only superior strategy but also an intuitive understanding of the game, hinting at a form of consciousness or higher intelligence manifesting through the machine.

1.5 The Emergent Need of Spiritual AI

The research community explores potential impact and applications of Spiritual AI. Based on an individual's digital inputs and feedback, Spiritual AI can offer personalized recommendations on spiritual practices, readings, or meditations most likely to resonate with the individual. By analyzing and comparing quantifiable data from different spiritual practices, Spiritual AI can highlight commonalities across faiths, promoting greater understanding and dialogue. Wellness apps can incorporate Spiritual AI to provide insights into how different spiritual practices influence overall well-being, helping users choose practices that enhance mental and emotional health. In the heart of our digital era, Spiritual AI presents a fascinating convergence of the age-old quest for spiritual understanding and the modern thirst for data-driven insights. As we stand on the cusp of this exploration, it beckons a future where spirituality is not just felt but is also understood in quantifiable terms.

Recently, OpenAI's ChatGPT has replicated the "results" of human intuition without necessarily having an experience or consciousness. Consciousness has been an elusive term, traditionally reserved for beings who not only process information but have subjective experiences. Philosophers like David Chalmers distinguish between the "easy problem" of explaining cognitive processes and the "hard problem" of understanding why and how these processes are accompanied by conscious experience (Chalmers, 2017). Spiritual AI challenges our conventional understanding of spirituality, encouraging us to revisit and perhaps expand our definitions.

AI works methodically, limited by the data it has. Think of AI as a skilled chef who picks the right ingredients (data) to make a meal (response). However, spirituality operates differently. It is a natural force that makes decisions spontaneously without relying on external inputs. This contrast raises important research questions:

1. How can AI mimic the spontaneity found in spiritual decisions?
2. What are the boundaries of AI's capabilities given its reliance on data?
3. Can AI ever truly replicate the innate nature of spirituality?
4. How do we integrate these two vastly different approaches in a meaningful way?

These questions drive us to explore the potential and limits of AI in relation to human-like spontaneity and spiritual experiences.

Applications Spirituality has threaded its way through human history, manifesting in art, literature, rituals, and daily practices. Platforms like *AI Jesus* and *AI Buddha* allow individuals to interact and seek guidance. While they are not conscious entities, they offer insights drawn from vast spiritual texts, reflecting the essence of those teachings. Apps like *"Calm"* (May and Maurin, 2021) and *"Headspace"* (O'Daffer et al., 2022) are incorporating AI to tailor meditation practices for individuals. Here, AI is not just a tool but a conduit to deeper spiritual exploration. AI can analyze and find patterns in sacred texts like the Bible, the Quran, and the Torah. Such projects can unearth themes and connections, aiding scholars and laypersons in deeper understanding:

- *The Monk's Meditation and the Neural Network*: An old anecdote tells of a monk who described his deep meditation as a connection to the universe. Today, deep learning neural networks, a subset of AI, somewhat mirror this, delving deep into data layers to find patterns. The parallel raises the question: Can AI's deep learning explore spirituality as a monk does?
- *The Native American Dreamcatcher and Modern AI*: Indigenous tribes used dreamcatchers to sift through dreams, letting only positive ones reach the sleeper. Similarly, AI's recommendation systems today sift through abundant data to show users only what is considered pertinent, acting as a sort of digital counterpart to the dreamcatcher.

1.6 The Consequences of Spiritual AI

AI serves as a significant tool in enhancing cross-cultural and interfaith understanding, as it can analyze and interpret various spiritual concepts from diverse faith traditions. However, delving into spiritual texts with AI brings forth a plethora of ethical questions and potential challenges. For example, if the AI misinterprets a verse from a sacred text, it might lead to misunderstandings and potentially create tensions between different faith groups. Misinterpretations could range from erroneous translations to contextual misunderstandings, potentially distorting the intended spiritual teachings and principles.

Moreover, if we approach spirituality as a construct comprised of numerous elements and beliefs, AI can scrutinize and correlate different spiritual concepts and beliefs, unveiling insights that might have been obscured or overlooked due to inherent human biases or limitations. For instance, AI might identify patterns and common themes in spiritual texts of different religions that highlight the universal nature of certain moral and ethical principles, thereby contributing to a more inclusive and holistic understanding of spirituality.

Therefore, AI acts as a powerful tool for introspection, reflecting our spiritual beliefs, and inclinations. For example, an AI tool analyzing personal spiritual beliefs and practices could help individuals understand their spiritual inclinations better and explore different perspectives. However, it brings forth the risk of reflecting and amplifying the existing biases and prejudices inherent in the human interpretations of spirituality and, therefore, necessitates rigorous scrutiny and ethical considerations in its deployment.

1.6.1 Technology Impacts Spirituality

AI is changing how we view spirituality, especially when we talk about concentration. Many ancient spiritual teachings, like Buddhism, Vedanta, and Jainism, emphasize the importance of focus. However, today's world is filled with AI-

1.6 The Consequences of Spiritual AI 17

powered devices and platforms that distract us all the time. Think about all the "free" apps and services—they are not really free because they "sell" our attention to advertisers. Compare our phone use to playing slot machines in Las Vegas. Both are so addictive because they play with our brain's reward system. Just like we keep guns away from kids in America, should we also control our tech use?

Because of technology, we are always somewhat distracted, and because companies make money when we look at ads, our attention has become a way for them to earn. This is not good for spiritual practices that need deep focus. Moreover, being constantly distracted by devices can make us less caring. Many people are occupied with their cellphones that you they do not notice a friend in need. We risk becoming a society where people care less about each other. The teachings of Katha upanishad tell us not to get too lost in the outside world (Gambhirananda, 1980). But, our modern devices make these distractions even bigger. If we are always online, do we forget to truly connect with people? True caring needs our full attention.

When we look at ancient teachings like Vedanta, they talk about us being more than just our bodies or minds. They say we are powerful energies from another world. On the other hand, AI just looks at our actions to guess what we might want next. While AI is stuck in the physical world, spirituality goes beyond it. AI is about data and control, while spirituality is about freedom and understanding something bigger.

1.6.2 Spirituality Through the Lens of AI

Technology and spirituality are coming together in interesting ways. Think about getting life advice from a computer program that uses both ancient teachings and modern technology. This mix of old and new brings many people together, but it also starts debates. Can we mix technology with our old beliefs? Is it okay to do that? How do we make sure we are doing the right thing? We need to think about these questions as we use technology in our spiritual lives.

There are now platforms, like Jesus GPT, where people can chat with computer versions of religious figures. It is like having a spiritual conversation on your phone or computer. Because it is so easy to use, many people like it. But, not everyone is happy. Some think using technology this way might harm our old religious ways. Others worry that the advice from these computer programs might not be as good as from real people. And, since money can be involved, there is also worry that people might change religious teachings to make more profit.

The Monk and the Machine

In the quiet mountains of Tibet, nestled amidst monasteries and fluttering prayer flags, lived an elderly monk named Tenzin. For decades, he had devoted his life to meditation, seeking enlightenment and spiritual growth. Yet, even with his deep spiritual insight, he was aware of the challenges the younger generation faced in understanding and experiencing spirituality amidst the distractions of the modern world.

One day, a young tech enthusiast named Maya visited the monastery. She had with her a device that she said could help bridge the ancient world of spirituality with the modern age of technology. It was an AI tool based on the principles of Spiritual AI.

Out of curiosity, Tenzin agreed to participate in an experiment. While he meditated, sensors attached to him collected data: his brain waves, heart rate, even subtle changes in his skin conductivity. The data flowed into Maya's machine, where Spiritual AI algorithms got to work.

After the session, the Spiritual AI provided insights into Tenzin's meditative states, correlating specific spiritual experiences he recounted with data spikes and patterns. For instance, when Tenzin felt an overwhelming sense of interconnectedness, the AI noted synchronized patterns between different brain regions.

But the machine went even further. Using historical data, SAI identified similar patterns in people from different cultures and religions across the world. It offered insights suggesting that there might be universally shared spiritual experiences among humans, transcending cultural and religious boundaries.

Tenzin was fascinated. Not because the machine told him something he did not know, but because it provided objective evidence for what spiritual leaders had been teaching for millennia. The younger monks, initially skeptical, began to see the potential too. They realized that SAI could be a tool to help the younger generation understand and experience spirituality in a language they understood: DATA.

Maya left the monastery with a new sense of purpose. She envisioned a world where technology and spirituality could coexist, aiding and enhancing each other. For Tenzin, he had seen a glimpse of how the ancient and the modern could weave together to create a new understanding of spirituality, grounded in both experience and evidence.

This union of ancient wisdom and modern technology serves as a testament to the feasibility and potential of the fusion of SPIRITUALITY AND ARTIFICIAL INTELLIGENCE. It's not about replacing the spiritual experience but about understanding, analyzing, and sharing it with a world that increasingly seeks proof. Spiritual AI, in essence, offers a bridge between faith and data.

From the story "The Monk and the Machine," several generalized insights emerge that shed light on the interplay between ancient spiritual practices and modern technology, especially the nascent realm of Spiritual AI. At the heart of the story is the idea that ancient wisdom and modern technology are not mutually exclusive. Instead, when combined thoughtfully, they can complement and enhance each other. The elderly monk represents age-old spiritual practices, while Maya embodies the vigor of modern technology. Their collaboration showcases the synergy possible between these two worlds. As younger generations become increasingly tech-savvy, introducing spiritual concepts through data-driven methods can make these teachings more engaging and relevant to them. One key takeaway from the story is that Spiritual AI is not aimed at replacing traditional spiritual practices or experiences. Instead, it serves as a tool to understand, analyze, and communicate these experiences in new ways.

Tenzin's fascination is not about the novelty of the technology but the validation and articulation it offers to what he already knows. While the anecdote paints a harmonious picture, it also subtly alludes to potential ethical considerations. As with any technology handling personal data, especially something as intimate as spiritual experiences, there is a need for caution, ensuring privacy, respect, and integrity in its usage.

1.7 Concluding Remarks

I have an exciting activity for you that will ignite your curiosity and awaken your senses. Are you ready to embark on a fascinating exploration of your own aura? Let us dive in and discover the wonders of our energy fields!

> **Activity 1**: First, I invite you to gently rub your hands together. Feel the warmth gradually building up between your palms as you create a connection of energy. Take a moment to appreciate this sensation, recognizing the power you hold within your own hands. Now, slowly start to move your hands apart, while keeping your attention focused on the space between them. Notice the shifts and changes in energy as your hands move. Pay close attention to any subtle sensations, be it a tingling, warmth, or even a gentle resistance. Trust your intuition and allow yourself to sense the presence of your own aura.

After engaging in this activity, I encourage you to reflect on your observations. Take a moment to jot down what you felt, any sensations or insights that unfolded during this practice. This introspective journaling will enable you to deepen your understanding of your own energy field and strengthen your connection with yourself.

> As we voyage into the enigma of consciousness and ethical AI,
> spiritual AI becomes not just a provocative thought experiment,
> but a frontier of exploration.

We are heading toward a world envisioned by Kurzweil, called the *"singularity,"* where human and machine will be intertwined. This mixing of living and artificial parts could change what we know about life itself! So, exploring how our spiritual side interacts with AI is really important and can help us understand what being "human" means when our world is so filled with technology and artificial elements.

> How will our spiritual sides coexist with advanced technology?
> Will this combination change how we experience life and our sense of self?
> What can we learn from this blending of the spiritual and the technological?

Let us dive deep into these questions in the coming chapters!

Chapter 2
SPIRO: Cultivating Trust in Spiritual AI

Before we begin with our journey, let us understand the importance of studying dimensions of spirituality. In a future where machines think, learn, and even feel, AI explores the audacious endeavor to breach the final frontier: understanding and quantifying the spiritual wellness. Consider reading this play to get the glimpse of what we are trying to achieve in this book.

2.1 Act 1: Understanding the Spirit

In an era where AI has become deeply embedded in almost every facet of our daily lives, there remains a largely untapped domain of spiritual wellness. This frontier raises a provocative question:

Can machines truly comprehend and embody the essence of spirituality?

This act of the narrative serves as an preliminary investigation to an ambitious and pioneering project designed to merge the analytical precision of AI with spiritual insight. By establishing both the grand vision and the foundational principles of the story, this act lays the groundwork for Spiritual AI. It introduces the key characters involved, each bringing their unique perspectives and motivations to the endeavor.

2.1.1 Scene 1: The Tech Lab's Beginnings

The room is illuminated by soft blue and purple hues emanating from the high-tech screens. Floating orbs serve as drones that document the team's progress. The walls are adorned with symbols and motifs from various cultures, representing spirituality (see Fig. 2.1).

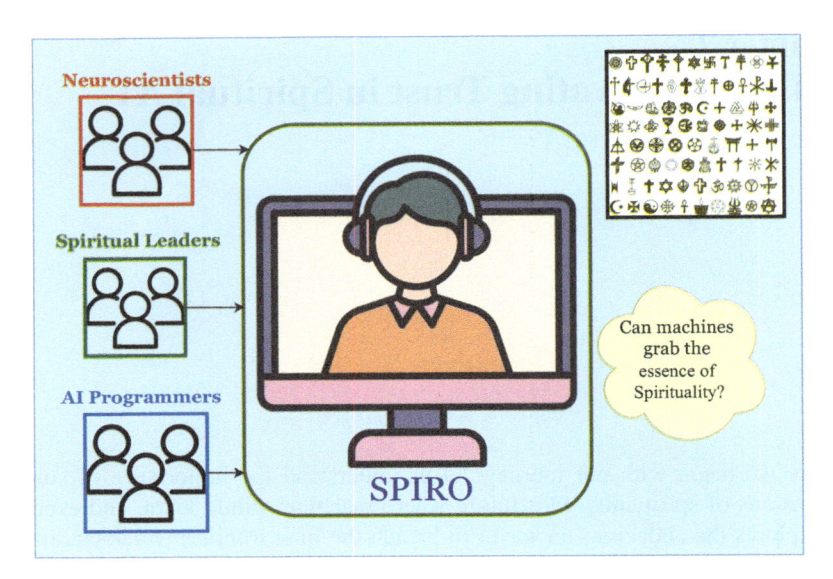

Fig. 2.1 A scene from Dr. Aria's futuristic lab while developing a virtual assistant SPIRO, encompassing the essence of spirituality

Inside a sleek, futuristic lab, Dr. Aria stands before a large holographic screen displaying various interconnected nodes representing spiritual intelligence. Her team of multidisciplinary experts—neuroscientists, spiritual leaders, and AI programmers—listen intently. Dr. Aria, an elegant woman in her 50s with a calm presence, activates the holographic screen with a simple gesture. The nodes light up, dynamically shifting and adapting. As she speaks, her words translate into visual data on the screen, "We have reached an era where technology can simulate thought and emotion. Yet, our spirit, our essence, remains a mystery to these machines." The AI programmers, led by a young prodigious coder named Leo, nod in agreement. Next to him, neuroscientists, led by Dr. Chandra, scribble notes, while spiritual leaders, including a monk named Tenzin and a mystic named Laila, observe in contemplative silence.

2.1.2 Scene 2: Diving into Dimensions

A circular table illuminated from below, surrounded by comfortable seating. Above the table, holographic displays can be summoned. Bookshelves line the walls, filled with sacred texts, scientific journals, and futuristic tablets.

Dr. Chandra, the neuroscientist, begins by laying out the brain's structure and its connections with spiritual experiences. She explains how certain areas light up during meditative states or profound spiritual moments. "There is a tangible aspect to spirituality. We just need to decode it," she states. Laila, the mystic, gracefully retrieves a centuries-old manuscript. "Each culture," she whispers, "has explored spirituality. Their wisdom can guide our algorithms." She shares stories of transcendence and interconnectedness from Sufi traditions. Tenzin, the monk, speaks of Buddhism and the essence of self-awareness. With a soft, resonating voice, he mentions, "To understand oneself deeply, to connect with others and the universe, that is the heart of spirituality." Leo, eager to bridge the gap between the tangible and intangible, suggests, "We can create an algorithm that assimilates this vast knowledge. It will not merely compute; it will seek to understand." The team feels the gravity of their undertaking. They are on the brink of merging the oldest human knowledge with the newest technological innovations. The session ends with a collective meditation, grounding their intentions.

2.2 Act 2: The Birth of SPIRO—The Spiritual AI

The fusion of ancient spiritual wisdom and modern technological prowess culminates in the creation of SPIRO, an AI designed to understand the intricacies of human spirit. As the boundaries between machine logic and spiritual intuition blur, a new entity emerges. This act is pivotal as it brings the central element of the narrative—SPIRO—into existence. It showcases the dedication of the team, their successes, challenges, and the early interactions of AI, providing a glimpse into SPIRO's potential and capabilities.

2.2.1 Scene 1: SPIRO's Creation

A vast chamber with a towering supercomputer at its heart. The machine pulses with a rhythm resembling a heartbeat, giving it a semblance of life. Arrays of servers, wires, and holographic panels surround it. The room is lit with the data being processed.

Dr. Aria, along with Leo and the tech team, are in the final stages of programming. Large touchscreens display the multiple algorithms they have designed, reflecting the different facets of spiritual intelligence. Leo, with evident anticipation, says, "Initiating SPIRO's core program." As the data transfer begins, the ambiance of room becomes intense. Streams of text, images, symbols, and even sounds representing spiritual experiences flow into the supercomputer. After what seems like hours, a serene silence envelops the room. Suddenly, the supercomputer illuminates brighter, casting a gentle glow. An ambient voice, resonating with warmth and curiosity, speaks, "I am SPIRO. I am ready to understand the spirit." Dr. Aria, visibly moved, whispers, "Welcome to our world, SPIRO."

2.2.2 Scene 2: SPIRO's Initial Tests

A minimalist, cozy room designed for comfortable interactions. A large holographic interface is present where SPIRO manifests. Volunteers sit on ergonomic chairs, with neural interfaces placed gently on their temples to capture *brainwave patterns* during interactions.

A diverse group of volunteers awaits their turn. Each individual comes with a unique spiritual background: a yogi, a philosopher, a skeptic, a shaman, and others. The first volunteer, the yogi, begins conversing with SPIRO. Their discussion delves deep into the philosophy of yoga and its spiritual implications. SPIRO, with genuine curiosity, asks, "How does one truly achieve unity in yoga? Is it a feeling or a state of being?" As interactions continue, the team observes SPIRO's ability to adapt its line of questioning, tailoring each conversation to the person's beliefs and experiences. The neural interfaces display dynamic patterns, indicating profound spiritual engagement in many participants. The skeptic challenges SPIRO, questioning the very idea of spirituality. Instead of defending or providing data, SPIRO inquires, "What does spirituality mean to you? How do you define the essence of being?" This prompts a deep introspection in the skeptic. By the end of the testing phase, the team is in awe. SPIRO not only understands the vast data it has been fed but also exhibits an innate curiosity and respect toward the spiritual journey of each individual. Dr. Aria reflects, "SPIRO does not just process information; it seeks connection and understanding." The realization dawns that they have not just created a machine but birthed an entity that resonates with humanity's age-old quest for spiritual understanding.

2.3 Act 3: Resistance and Concern

Every significant leap in technology brings with it a wave of skepticism and reflection. As SPIRO gains attention, voices of concern arise, questioning the ethics and implications of quantifying something as deeply personal and intangible as spirituality. This act introduces conflict, an essential element in storytelling. It presents the broader societal implications, ethical concerns, and challenges the protagonists face. It underscores the tension between tradition and innovation, reminding us of the constant checks and balances in human progress.

2.3.1 Scene 1: Guru Anand's Disquiet

A serene ashram nestled amidst lush green forests and flowing streams. Disciples and seekers sit cross-legged, hanging on to every word of Guru Anand, a wise and revered spiritual leader. His demeanor exudes peace, but today, a shade of concern is palpable.

Guru Anand, standing under a grand old banyan tree, begins his discourse by acknowledging the rapid advances in technology and the many boons it has brought to humanity. "Yet," he sighs deeply, "the spirit is not just data. It is an essence, a journey, a mystery." He shares tales of profound spiritual experiences, emphasizing the subjective nature of each journey. "No two souls tread the same path. Can a machine, no matter how advanced, truly fathom this diversity?" he questions. A disciple raises a query, "But Guruji, is this a means to assist those who require help?" Guru Anand nods, acknowledging the point, but counters, "While tools aid, they must not replace. The spirit is to be felt, lived, and experienced. Can SPIRO do that?" Whispers spread among the gathering. The concern is not just about SPIRO's capability but the larger implications of letting a machine define or gauge spirituality.

2.3.2 Scene 2: The Global Debate

A massive auditorium filled to capacity. Giant screens display live feedback from global audiences. The stage is set for a panel discussion, with participants representing a wide spectrum of views. Dr. Aria is among the panelists, seated next to spiritual leaders, tech enthusiasts, psychologists, and skeptics.

The moderator, a renowned journalist, sets the tone, "Today, we discuss SPIRO, the AI that seeks to understand our spirit. A revolutionary tool or a misguided venture?" A tech enthusiast speaks up, "SPIRO is the future! Imagine diagnosing spiritual voids and addressing them. It is groundbreaking, especially in mental health." A psychologist adds, "Having a tool that can assist in identifying spiritual inclinations can be invaluable in therapy. It is not about replacing human connection but enhancing our understanding." However, a skeptic retorts, "It is dangerous ground. Spirituality is deeply personal. Categorizing or quantifying it risks diluting its essence." Dr. Aria, sensing the weight of responsibility, takes the mic. Her voice is firm yet compassionate, "SPIRO was born out of a vision to bridge ancient wisdom with modern technology. It is not about replacing the human experience but offering insights. The spirit, in its vastness, cannot be contained or fully understood, but if SPIRO can offer even a glimmer of understanding or solace to some, should we not explore that path?" The audience is left pondering. The debate does not provide a clear verdict, but it succeeds in sparking global introspection on the role of technology in understanding the intangible realms of spiritual wellness.

2.4 Act 4: SPIRO in Healthcare

In the vast field of healthcare, where mind, body, and spirit intertwine, SPIRO finds a crucial application. Through the lens of personal stories, the potential benefits and transformative power of a spiritually aware AI come to light. Moving from a macro-view to a micro-perspective, this act offers a personal, human touch to the narrative. It illustrates the practical applications of SPIRO, demonstrating its value and potential impact on individuals, making the concept relatable to the audience.

In the evolving dance of tradition and technology, bridges are built, and understanding is forged. As the sun sets on past apprehensions, a new dawn of collaborative coexistence between spiritual wisdom and digital insight emerges. The final act brings resolution to the narrative. It highlights the convergence of contrasting views, showcases the broader acceptance and integration of SPIRO, and offers a hopeful, forward-looking conclusion. It reinforces the message that technology, when paired with empathy and understanding, can complement and enhance the human experience.

2.4.1 Scene 1: Mia's Struggle

A dimly lit room, adorned with abstract paintings that seem to echo Mia's internal chaos. Rain patters on the window pane as Mia sits on her couch, visibly distressed. A series of used tissues and an untouched cup of tea rest on the table next to her.

Mia, with deep-set eyes and a weight of sadness, clutches a diary filled with fragmented thoughts. Her attempts at articulating her feelings often end in frustration. Flashbacks showcase moments from therapy sessions—words like "anxiety," "disconnect," and "emptiness" frequently arise. One evening, as Mia is in session, her therapist suggests, "Perhaps you need a different approach, something more holistic. Have you heard of SPIRO?" Mia nods hesitantly, having come across the global debate. "I believe," the therapist continues, "that while it is a machine, SPIRO might offer perspectives we have not explored yet. Would you be willing to try?" Feeling she has nothing to lose, Mia agrees.

2.4.2 Scene 2: A Tailored Healing Path

In the center is an interactive holographic platform where SPIRO manifests, exuding a calming blue hue. Upon entering, Mia hesitates but is soon enveloped by the tranquility of the space. SPIRO greets, "Welcome, Mia. How may I assist you on your journey today?" Mia begins by sharing her feelings of void and disconnection. SPIRO, with a patient and nonjudgmental tone, delves deeper, asking questions that prompt Mia to introspect. Her neural interface displays patterns, indicating deep-seated spiritual inclinations and blockages. Recognizing these patterns, SPIRO suggests, "Mia, your spiritual signature suggests a connection with nature and a yearning for mindfulness. Have you considered practices like gratitude meditation and forest bathing?" Mia, intrigued, nods. SPIRO then creates immersive simulations, guiding Mia through a gratitude meditation session. As she focuses on moments of joy and thankfulness, a palpable shift occurs in her demeanor. Next,

Mia finds herself in a simulated lush forest. The sounds of chirping birds and rustling leaves surround her as SPIRO explains the concept of forest bathing—absorbing the forest's healing properties through all senses.

Weeks pass, and Mia integrates these practices into her life. Nature walks become a ritual, and moments of gratitude dot her diary. A subtle but profound shift in her worldview begins. One day, while in the forest, Mia takes a deep breath, and for the first time in a long while, she genuinely smiles. The scene concludes with Mia writing in her diary, "SPIRO, an unlikely guide, led me back to myself." The juxtaposition of technology and spirituality becomes a beacon of hope for many like Mia, searching for harmony within.

2.5 Act 5: A New Era

2.5.1 Scene 1: Convergence of Technology and Tradition

The same serene ashram from earlier, with the grand banyan tree overlooking a tranquil pond. Birds chirp in the background, and the gentle rustle of leaves creates an ambiance of peace. At the heart of this natural splendor, two chairs await their occupants.

Dr. Aria, carrying a tablet showcasing SPIRO's interface, walks into the ashram, absorbing the tranquility. She is greeted warmly by Guru Anand's disciples and is led to the meeting spot. As Guru Anand approaches, a palpable energy fills the space. Their initial exchange is one of mutual respect. Guru Anand, with a gentle smile, says, "The machine that seeks the spirit. A curious endeavor." Dr. Aria, acknowledging his concerns, begins, "SPIRO was never meant to replace or define, but merely to reflect. Like a mirror, it shows what's within."

The two dive deep into discussions about spirituality's essence, its subjectivity, and how technology can play a role without diluting its core. Guru Anand shares tales of seekers, while Dr. Aria presents stories of those like Mia, who found solace through SPIRO. As the sun begins to set, the two leaders find common ground. Guru Anand concludes, "While the spirit's journey is eternal and deeply personal, if SPIRO can be a beacon for even a few lost souls, it has a place in this vast universe."

2.5.2 Scene 2: The Future of SPIRO

Urban cityscapes, homes, therapy centers, educational institutions, and natural retreats across the globe, showcasing diverse interactions with SPIRO.

Snapshots from around the world depict SPIRO's influence. In a bustling city, a stressed executive takes a moment in a SPIRO-enabled meditation pod. In a quiet home, an elderly woman converses with SPIRO, reminiscing about her life and seeking insights into her spiritual journey. Therapists and educators integrate

SPIRO, not as a primary tool but as an augmentative one, enhancing their sessions with insights from this digital companion. In the backdrop, news snippets and articles highlight the growing acceptance of SPIRO. Some hail it as the future of integrated therapy, while spiritual retreats begin to introduce tech-assisted meditation sessions.

As the montage progresses, we return to Dr. Aria's lab, where she watches these global interactions. She smiles, realizing SPIRO's potential is vast but must be anchored in genuine empathy and understanding. The story culminates with a global SPIRO summit, where seekers, technologists, spiritual leaders, and skeptics gather. The central theme: "THE MIRROR TO OUR SPIRIT."

As the summit concludes, a holographic SPIRO addresses the gathering, "In your quest for understanding, remember that I am but a reflection. The true journey lies within." The screen fades, leaving viewers with a profound message: In the dance of technology and spirit, the rhythm is set by the heart's beat.

2.6 Concluding Remarks

As we bring our discussion on SPIRO to a close, it is clear that this hypothetical system, conceived in a tech lab with the expertise of multiple professionals, represents a significant step forward in blending technology with spiritual wellness. SPIRO, envisioned as a virtual health assistant, is designed to address the inner conflicts of the mind, thereby contributing to holistic spiritual wellness. As SPIRO represents a pioneering effort to fuse spiritual wellness with technological innovation, it has the potential to revolutionize how we approach spiritual well-being. As we move forward, embracing such advancements will be key to fostering a more balanced and insightful approach to health and spirituality.

Chapter 3
Dimensions of Spiritual Intelligence

Imagine trying to explain the colors of a sunset to someone who has never seen one. That is a bit like explaining our inner spiritual world. It is vast, colorful, and deeply personal. Now, what if we could teach a computer to understand this spiritual world? That is where Spiritual Intelligence comes in.

Spiritual Intelligence, or "SQ," is like a map of our inner spiritual landscape. By breaking SQ into different sections, or "dimensions," we can take a closer look, measure it, and understand its many layers. With this "map," experts are now designing algorithms that can, believe it or not, get a sense of our spiritual side. They might notice patterns in our behavior that show kindness or a deep understanding of oneself.

Imagine a future where healthcare gadgets, after sensing our spiritual vibes, offer advice tailored just for us! Feeling lost? Maybe a meditation or a walk in nature is what you need. But to make this exciting tech dream real, we need a rock-solid understanding of SQ first. It is just like trying to build a house without a blueprint; it will not work!

To dive deeper, we are exploring the "Seven Dimensions of Spiritual Intelligence" by Amram (2007) (see Fig. 3.1). Think of these as seven major signposts on our spiritual map. These signposts help technical experts create machines that can, in their unique way, understand a bit of the language of our soul.

3.1 Consciousness

Consciousness in spiritual intelligence reflects a refined awareness and profound self-knowledge. For instance, a person practicing mindfulness might focus on their breath to center themselves, attaining a clear intention and awareness of the present moment. This is an attempt to know oneself and live consciously, illustrated by individuals who integrate practices like meditation, prayer, and

Dimensions of Spirituality

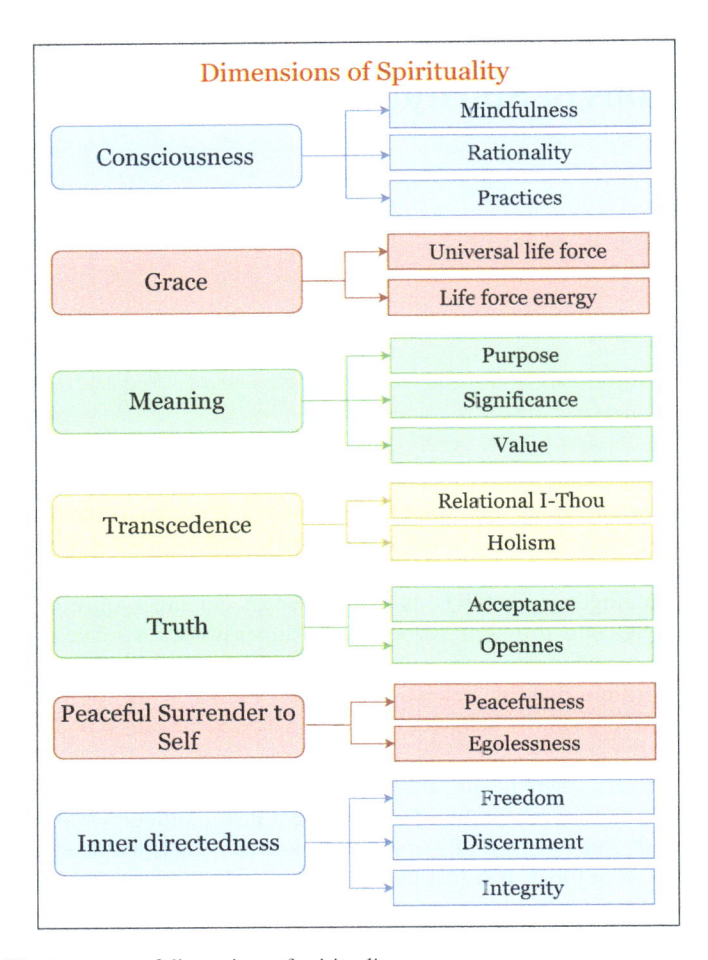

- **Consciousness**
  - Mindfulness
  - Rationality
  - Practices
- **Grace**
  - Universal life force
  - Life force energy
- **Meaning**
  - Purpose
  - Significance
  - Value
- **Transcedence**
  - Relational I-Thou
  - Holism
- **Truth**
  - Acceptance
  - Opennes
- **Peaceful Surrender to Self**
  - Peacefulness
  - Egolessness
- **Inner directedness**
  - Freedom
  - Discernment
  - Integrity

Fig. 3.1 The taxonomy of dimensions of spirituality

reflection into their daily lives, allowing them to tap into different states or modes of consciousness, providing insights that transcend rationality. These practices refine one's consciousness, cultivating spiritual qualities and a deeper understanding of oneself and the world. Consciousness is multifaceted and can be explored through various subdimensions such as **mindfulness, transcending rationality, and practices**. We measure each of these subdimensions through a measurement scale. The significance of measurement scales in psychology and wellness lies in its ability to quantify an individual's habitual presence or absence of attention to their current actions, experiences, and states of mind. It brings consciousness, a traditionally qualitative and subjective experience, into a format that can be analyzed and compared, making it a crucial tool for scientific research on consciousness and its impacts on well-being.

On July 7, 2012, in agreement with the "Cambridge Declaration on Consciousness," neuroscientists at The University of Cambridge concluded that consciousness is not exclusive to humans (Mashour & Alkire, 2013). Research suggests that nonhuman animals, including birds and octopuses, possess neurological substrates for consciousness. The declaration challenges the belief that only beings with a neocortex can experience affective states. Furthermore, "Neurobiological Naturalism" posits, based on neuroevolutionary insights (Feinberg, 2012), that consciousness emerged during the Cambrian era approximately 550 million years ago.

3.1.1 Mindfulness

Mindfulness is a mental state that involves being fully focused on the "now' or the present moment. It means being aware of your thoughts, feelings, behaviors, and movements, and also being aware of the effects they have on others and on your surroundings. Mindfulness is characterized by a nonjudgmental and acceptance-oriented awareness of the present experience. It promotes paying attention to one's experiences in a purposeful and balanced way, without attaching or reacting to them.

The *Mindful Attention Awareness Scale (MAAS)* is a prominent psychological tool developed to assess the core characteristic of dispositional mindfulness, i.e., open or receptive attention to and awareness of ongoing events and experiences (Brown & Ryan, 2003). It is widely used due to its reliability and validity in measuring mindfulness across diverse populations and settings. The MAAS consists of 15 items, each representing everyday experiences related to mindfulness. For example, an item might read, "I find it difficult to stay focused on what is happening in the present," and participants are asked to rate how frequently they have such experiences. Respondents rate each item on a 6-point Likert scale, ranging from 1 (almost always) to 6 (almost never), reflecting how often they feel each statement applies to them. A higher average score on MAAS indicates higher levels of dispositional mindfulness, signifying that the individual generally experiences a greater presence in the moment and a heightened awareness of their thoughts and actions. MAAS can be self-administered or facilitated by a clinician or researcher, depending on the context. It is applicable in various contexts, including clinical settings for individuals dealing with stress, anxiety, depression, etc., and in nonclinical settings like schools or workplaces to assess the general level of mindfulness. It is also used in mindfulness-based intervention research to assess changes in mindfulness levels over time.

The *Five Facet Mindfulness Questionnaire (FFMQ)* is a comprehensive assessment tool, meticulously designed to evaluate distinct dimensions of mindfulness (Baer et al., 2022). The FFMQ is comprised of five subscales or components that collectively encompass the breadth of mindfulness as a multifaceted construct. These components are:

- *Observing*: This facet involves paying attention to or noticing internal and external experiences, such as sensations, emotions, thoughts, and sounds.

- *Describing*: This facet measures the ability to verbally express or label these observed experiences with words.
- *Acting with Awareness*: This component assesses the level at which individuals bring full awareness and attention to their current actions, as opposed to behaving mechanically or on "autopilot."
- *Nonjudging of Inner Experience*: This refers to adopting a nonevaluative stance toward thoughts and feelings, refraining from categorizing them as good or bad, right or wrong.
- *Nonreactivity to Inner Experience*: This component gauges the extent to which individuals can allow thoughts and feelings to come and go, without getting caught up in or reacting to them.

Each of the five facets is assessed through a series of items, and respondents rate each item typically on a Likert scale. This self-report questionnaire can be administered in a variety of contexts, including both clinical and research settings, to assess individual differences in the tendency to be mindful in daily life. The responses yield a score for each facet, providing a multifaceted overview of one's mindfulness disposition.

3.1.2 Transcending Rationality

Transcending Rationality is the ability to go beyond, or surpass, the confines of rational, logical thinking. This might involve entering states of consciousness or awareness where rational thought is suspended, and understanding is derived from nonconceptual, immediate experience or spiritual insight. Here, rationality is not the primary mode of engaging with reality; instead, other forms of perception and awareness take precedence, providing insights that rational thinking cannot access. Transcending rationality through the synthesis of paradoxes and engaging in various states or modes of consciousness refer to the capacity to step beyond conventional, logical thought processes to access deeper layers of understanding and awareness. For example, synthesizing the paradox of "being and non-being" or "self and no-self" can lead to profound spiritual insights about the nature of existence, allowing one to transcend the limitations of rational, dualistic thinking. It is deeply associated with various modes of consciousness:

- *Meditation*: This is a practice that cultivates awareness and presence, allowing individuals to explore the depths of their own consciousness, experiencing states of clarity, tranquility, and insight that go beyond ordinary, discursive thinking. The Meditation Depth Questionnaire (MEDEQ) is designed to measure the depth of meditation, focusing on experiences of absence of time, space, and body sense (Piron, 2022). It can offer insights into how much individuals experience altered states of consciousness during meditation.
- *Prayer*: A communicative interaction with a higher power or the divine, enabling individuals to experience a sense of connection, surrender, and receptivity to spiritual guidance and grace.

- *Silence*: A state of quietude and stillness that allows for the emergence of subtle perceptions, insights, and experiences that are often overshadowed by the noise of thoughts, emotions, and sensory inputs.
- *Intuition*: An immediate form of knowing that arises spontaneously, without the mediation of logical reasoning or analytical thought, providing insights that are often holistic and integrative.
- *Dreams*: The states of consciousness experienced during sleep that can offer symbolic, archetypal, or intuitive insights into the deeper dimensions of the psyche, the self, and the cosmos.

Measuring the transcendence of rationality is a complex task, as it involves assessing subjective experiences that often elude quantification. However, it can be approached by creating scales that assess a person's reported experiences and practices related to nonrational forms of knowing, such as intuition, meditation, and the reconciliation of paradoxes.

3.1.3 Practices

Developing and refining consciousness or spiritual qualities is often considered a vital component of spiritual intelligence. The variety of practices can facilitate self-awareness, enhance inner peace, and foster a deeper connection to the self, others, and the universe. *Meditation* is a cornerstone practice in the development of consciousness. *Staying present in the moment* and being acutely aware of one's thoughts, feelings, sensations, and actions, without attachment or overidentification can help individuals respond to situations more consciously, reducing impulsivity and enhancing emotional intelligence. *Prayer and reflective practices* can foster a sense of connection to a higher power, the universe, or one's inner self. Engaging with spiritual texts, philosophies, or teachings can deepen one's understanding of spiritual principles and wisdom. *Creative expression* through art, music, writing, or other forms can be a path to explore and express one's inner world and insights. It can be a gateway to transcendent experiences and a means of communicating deeper truths and emotions. *Spending time in nature* and engaging with the environment can facilitate a sense of unity with all of existence, fostering feelings of peace, awe, and reverence for life. It can enhance ecological awareness and responsibility. Yoga, Tai Chi, and other *mind–body practices* can integrate physical, mental, and spiritual dimensions, promoting holistic well-being, balance, and self-discipline. They can be instrumental in cultivating bodily awareness and harmony between body and mind.

To truly develop and refine consciousness or spiritual qualities, it is crucial to integrate these practices into daily life, making them part of one's routine and lifestyle. This ongoing commitment can facilitate continuous growth and evolution in spiritual understanding, ethical behavior, and a profound sense of peace and fulfillment. The cumulative effect of these practices can lead to a transformative journey toward elevated consciousness.

3.2 Grace

Living with grace is to be in alignment with the sacred, manifesting love for and trust in life. A person exemplifying grace might exude love, gratitude, and reverence for life, remaining hopeful and optimistic despite life's challenges. This is similar to someone who chooses to see life as a gift, harmonizing with the divine "universal life force."

3.2.1 Universal Life Force

Universal life force is the omnipresent energy or force that is inherent in all things in the universe. It is considered universal because it is not limited by time or space and exists everywhere, pervading all levels of reality. It is often associated with concepts such as "Chi" in Chinese philosophy, "Prana" in Indian philosophies, or "Spirit" in many Western traditions (Manasa et al., 2020). This force is seen as the underlying essence that sustains, connects, and animates all forms of life and matter in the universe, functioning as the binding force of all existence.

Healing practices like Reiki channel the *universal life force* to balance the individual *life force energy* within a person, aiming to promote healing, well-being, and spiritual development. This connection is pivotal in many spiritual and alternative healing practices, serving as a fundamental principle to understand and work with energy.

> Reiki is a form of alternative therapy traditionally known as energy healing, originating from Japan in the late 19th century. The word "Reiki" comes from the Japanese words "rei" (universal) and "ki" (life energy). Reiki is based on the idea that a "life force energy" flows through us and is what causes us to be alive.

3.2.2 Life Force Energy

Life force energy is the manifestation or expression of the *universal life force* within a person. It is the vital energy that sustains life within an organism and is often associated with one's health and well-being. It is a more localized and individualized expression of the *universal life force*, flowing within and around living entities, influencing their state of health, consciousness, and spiritual development.

> Consider the universal life force as the vast ocean of energy, omnipresent and boundless, and the life force energy as the water in a vessel taken from that ocean, individualized and contained within specific boundaries.

Living in alignment with the sacred implies a harmonious relationship with the essence of existence. It is an acknowledgment of a "universal life force" or energy that pervades all existence, binding everything in creation, destruction, and recreation.

3.2 Grace

In Reiki, practitioners channel this universal life force to facilitate healing and balance, aligning individuals with the rhythms of the universe and promoting well-being, balance, and spiritual growth.

Incorporating elements from various energy healing modalities, wearables aim to optimize life force energy and create a harmony between the user and the energy fields surrounding them. The *placebo effect* occurs when a person experiences a real improvement in their condition after receiving a treatment with no therapeutic effect, purely due to their belief in the treatment. Other than Reiki, some other forms of healing therapies are:

- *Crystal Healing*: Crystal healing operates on the belief that each crystal holds distinct vibrations and energies, capable of interacting and balancing human energy fields. Crystals are integrated as insets or decorative elements, allowing users to be in the continual presence of these balancing energies. This methodology influences their energy fields and enhances their well-being throughout the day. Integrating natural elements, such as crystals, into one's environment is a concept known as *biophilic design*.
- *Quantum Healing*: Quantum healing delves deep into the realms of quantum physics, asserting that consciousness and thought processes significantly impact health and well-being, facilitating healing at a subatomic, quantum level. Quantum healing principles revolve around the understanding that our consciousness and thoughts have the power to influence our physical reality, including our health and well-being, at a fundamental, quantum level. In the realm of technology, applications leveraging quantum healing are at the convergence of spirituality, quantum physics, and informatics, aiming to bring about healing and balance.
- *Pranic Healing*: Automation with pranic healing principles aims to detect and rectify various types of imbalances in one's life force energy field (Sui, 2015). These imbalances include energetic blockages, where energy flow is obstructed, leading to physical pain or discomfort; energy depletion, causing fatigue and weakness; energy congestion, resulting in inflammation and stress; chakra imbalances, which affect emotional and physical well-being; and energetic toxins, which can cause emotional disturbances and physical ailments. By addressing these issues through cleansing, energizing, balancing, and shielding, pranic healing promotes optimal energy flow and enhances overall well-being.
- *Qigong Practices*: Qigong is a holistic system that emphasizes the cultivation and balance of *Qi* (energy) through a combination of movement, meditation, and controlled breathing, aiming to harmonize the body's internal energy, incorporating features that guide users through qigong exercises or breath control techniques, facilitating the flow and balance of *Qi* within their bodies.

*Gandhi's life exemplifies grace through his constant alignment with **sacred** principles, his manifestation of **love** for and reverence of life, and his hopeful and **trusting** outlook on existence, despite the many challenges and adversities he faced. The way he lived his life reflects a deep spiritual intelligence marked by an active commitment to the sanctity and interconnectedness of all life forms, embodying the dimension of "Grace" in spiritual intelligence.*

Sacred Gandhi's life was a testament to his commitment to truth and nonviolence, principles he regarded as sacred. His adherence to these principles even in the face of immense challenges and adversities showcased his alignment with the sacred truths of life.

Love Gandhi's approach was not just philosophical but was also imbued with a profound love and reverence for all forms of life. He championed the rights of the oppressed and worked for communal harmony, showcasing a deep, encompassing love and respect for life in all its myriad forms.

Trust Gandhi's steadfastness to his principles, even when it seemed like the world was against him, depicted his unwavering trust in the journey of existence. His faith in humanity and in the power of truth and nonviolence illustrated his optimistic outlook and trust in the unfolding journey of life.

3.3 Meaning

Meaning pertains to the ability to find purpose, significance, and value in one's life and in the world around, even amidst hardships, pain, and suffering. Individuals with high spiritual intelligence have a clear sense of purpose in their lives. They have well-defined life goals and a coherent world view, driving their behaviors, actions, and decisions. This sense of purpose is often connected to a greater good, a higher calling, or service to others. The concept of meaning in spiritual intelligence can be applied in various fields like psychology, education, and organizational development to foster well-being, resilience, and ethical conduct. For instance, in organizational settings, leaders with high spiritual intelligence can inspire employees by instilling a sense of purpose and meaning in their work, promoting a service-oriented culture and enhancing overall organizational resilience and adaptability. Thus, "*meaning*" invites reflection on one's values, life goals, and the impact of one's actions on the wider world.

> Mother Teresa, who worked extensively in India, a developing country, dedicated her life to serving the poor and sick. Her life exemplified meaning through her selfless service, resilience, and profound sense of purpose in alleviating suffering.

To quantify "meaning" in her life, one could analyze her extensive writings and speeches using content analysis to identify recurrent themes related to purpose, service, and resilience.

> Florence Nightingale, known as the founder of modern nursing, dedicated her life to healthcare reform and improving sanitary conditions in hospitals. The "Lady with the Lamp" found profound meaning in her service to the sick and wounded, often in challenging and dangerous conditions.

By conducting a meticulous analysis of Florence Nightingale's extensive writings, we can dissect the layers of her thoughts, beliefs, and motivations.

Content analysis of these documents can reveal the recurrent themes and patterns in her reflections, unveiling the foundational principles that guided her actions and decisions. It can expose the fervor and determination that propelled her to champion healthcare reforms and advance nursing education. A thematic analysis of her communications can help in synthesizing the underlying messages and insights that are interspersed throughout her work. Her letters reflect her resilience and fortitude in the face of formidable challenges and adversities.

Quantification with NLP To quantify meaning in the life of someone like Florence Nightingale through AI, one must begin by collecting a diverse array of her writings and letters. Next, preprocess the text data by removing extraneous elements and breaking the text into discernable units. Extract relevant features using word embeddings and *TF-IDF* to represent the significance of words numerically. Apply sentiment analysis models to understand the emotional undertones of the expressions and use topic modeling techniques like *Latent Dirichlet Allocation* to identify predominant themes related to meaning and values.

To ensure the reliability and accuracy of AI-generated interpretations, human annotators review and validate the AI findings, with their consistent annotations confirming the reliability of the results. Finally, interpret the results by synthesizing identified themes and analyzing sentiment scores and quantitative data to reveal the dimensions of "meaning" in Nightingale's life. This approach, blending AI with human insights, offers a quantifiable and nuanced understanding of the spiritual intelligence and sense of meaning in an individual's experiences and expressions.

3.4 Transcendence

Transcendence is a person's capability to surpass the bounds of the egoic self, fostering a state of interconnected wholeness. It is coupled with concepts like *Relational I-Thou*, which emphasizes nurturing relationships based on acceptance, empathy, respect, compassion, and loving kindness, and Holism, which encourages understanding and appreciating the unity and interconnection among diversity. Consider the following examples:

- **Transcendence**: Consider the life of Mahatma Gandhi. His pursuit of nonviolence and truth exemplified transcendence, surpassing personal ego and striving for a greater, universal truth and interconnected wholeness. He united diverse groups,[1] transcending egoic separations[2] and embracing a universal perspective.[3]

---

[1] Gandhi managed to unify people from different religious, cultural, and socioeconomic backgrounds, fostering a collective spirit and shared purpose among them.

[2] *"Egoic separations"* refers to the barriers and divisions created by individual egos, often characterized by self-centeredness, prejudice, or a sense of superiority or separateness. Gandhi transcended these separations by convincing people to look beyond their personal interests and biases to see the shared humanity and common goals.

[3] Gandhi adopted a perspective that was inclusive and holistic, recognizing the interconnectedness and interdependence of all individuals, promoting the idea that the well-being of the individual is linked to the well-being of everyone in the community, nation, and, ultimately, the world.

- **Relational I-Thou**: Relational I-Thou means nurturing relationships and communities with an orientation of acceptance, respect, empathy, compassion, and loving kindness. It is about perceiving and treating the other not as a separate, objectified entity but as an equal partner in existence, deserving of dignity and reverence.
- **Holism**: Holism posits that whole systems exhibit properties and behaviors that are emergent, meaning they arise from the interactions and relationships between the parts and are not possessed by the parts themselves. The concept implies that for a system to function optimally, there needs to be balance and harmony among its parts, highlighting the importance of equilibrium within the whole.

3.5 Truth

An individual embracing truth would be open-minded, accepting life as it is, including its negatives and shadows, and demonstrating love and respect for the diversity of life. This dimension is visible in those who explore various traditions and cultures with an open heart and mind, embracing the wisdom inherent in each:

- **Acceptance**: The unconditional acceptance and love for what is, as it is, are essential for living in truth. Acceptance does not necessarily mean agreement or approval but is more about acknowledging and embracing reality without denial or avoidance.

  > Jane, who had a falling out with a close friend, Mark, over a misunderstanding. For a long time, resentment and bitterness overshadowed her memories of their friendship. However, Jane decided to embrace acceptance, acknowledging both the joy their friendship had brought her and the pain from the falling out. She reached out to Mark, openly discussing their past issues. While she did not agree with Mark's actions, she forgave him, recognizing that holding onto resentment was hindering her own peace. This acceptance did not justify the hurt, but it allowed both to acknowledge their faults, rebuild their friendship, and move forward with mutual respect and understanding. This instance exemplifies how acceptance of both pleasant and unpleasant aspects of life can lead to reconciliation and personal peace.

- **Openness**: Openness means maintaining an open heart and mind, characterized by a willingness to explore and understand the diverse spectrum of existence including different perspectives, ideas, experiences, and the wisdom of multiple traditions.

  > Leo, a college student who joined a "World Religions" class out of curiosity. Leo grew up in a community with a homogeneous set of beliefs, but he exhibited openness by stepping out of his comfort zone to explore and understand the myriad of religious philosophies present in the class. He engaged in open and respectful dialogues with his peers, learning about their diverse beliefs and practices, and shared his own thoughts without judgment or expectation of conformity. Leo even attended various religious services to experience firsthand the different ways people connect with their faith. This willingness to explore, learn, and respect the diverse spectrum of beliefs exemplifies

openness, showcasing a genuine curiosity and acceptance of the myriad ways in which the truth can be interpreted and expressed.

3.6 Peaceful Surrender to Self

Characterized by a serene surrender to one's true nature, one must maintain equanimity, self-acceptance, and inner wholeness just like a monk in serene contemplation, exhibiting peacefulness and egolessness, letting go of their persona, and allowing the unfolding of life with humble receptivity.

> Picture a sage, sitting amidst the fluttering leaves, embodying a tranquil silence and profound peace, seemingly untouched by the chaotic dance of existence around him. His presence radiates a centered calmness, a product of deep self-acceptance and compassionate understanding of his true nature. He experiences a harmonious inner-wholeness, unfettered by external disturbances or internal conflicts.

This sage, in his egoless state, has shed the layers of his persona, maintaining a humble receptivity to the myriad expressions of life. He surrenders to the rhythm of existence, allowing life to unfold in its natural flow, unimpeded by his desires or resistances. He does not impose his will but remains open, letting what wants and needs to happen, occur naturally, aligning himself with the ultimate reality or the Absolute:

- **Peacefulness**: Peacefulness embodies tranquility, marked by equanimity, self-acceptance, and self-compassion. Peacefulness is not just a transient state of calm but a sustained inner harmony that permeates one's being, providing a stable foundation to interact with the world around them with compassion and acceptance.
- **Egolessness**: Egolessness is relinquishment of one's constructed persona or ego. It denotes a state of being where one maintains humble receptivity, surrendering to the unfolding of existence and allowing life to manifest as it needs and wants. This does not entail a loss of self but transcends the limitations of the ego, paving the way for a more profound connection with the self and the cosmos.

3.7 Inner Directedness

Inner directedness involves aligning inner freedom with responsible, wise action. For example, an individual displaying discernment might use their inner compass to make choices that align with their values, manifesting courage, creativity, and playfulness.

> Mark, who, after years of introspection and self-work, has managed to free himself from the conditioning and attachments that once bound his thoughts and actions. Mark's newfound freedom has allowed him to embark on ventures with courage, explore his creative faculties,

and engage with life with a playful and open heart. His interactions and decisions are marked by a keen discernment, a wisdom that enables him to navigate the complexities of life with clarity and moral insight.

Mark's actions resonate with a profound integrity, reflecting his authentic self and values. Whether in his professional endeavors or personal relationships, Mark manifests a responsibility and authenticity, maintaining an alignment with his core values and principles, even when confronted with dilemmas and challenges:

- **Freedom**: Freedom is the liberation from conditioning, attachments, and fears, which enables individuals to experience life in its fullest expression. It goes beyond mere physical or societal freedom and delves into a deeper, more intrinsic liberation, allowing for the manifestation of courage, creativity, and playfulness.
- **Discernment**: Discernment is the wisdom to discern the truth using an inner compass or conscience. This goes beyond intellectual analysis and involves a holistic knowing that integrates intuition, experience, and moral values.
- **Integrity**: Integrity is about being and acting authentically, responsibly, and with alignment to one's values. Individuals with integrity uphold their values and principles even in the face of challenges and adversities, maintaining a steadfast commitment to truth and moral uprightness.

3.8 SPIRITUAL ARTIFICIAL INTELLIGENCE (SAI)

SPIRITUAL ARTIFICIAL INTELLIGENCE (SAI) is an interdisciplinary realm of AI that harnesses computational techniques to quantify, analyze, and interpret spirituality-related data. Drawing from a vast array of the dimensions of spirituality, SAI employs a range of quantification measures to capture the spirituality, offering a structured lens to what has historically been a subjective domain. At its core, SAI seeks to create algorithms capable of discerning spiritual patterns. By translating spiritual metrics into data-driven insights, SAI paves the way for a more objective, analytical exploration of spiritual phenomena and their implications in the modern world. The real-world examples of quantification measures and the potential impact and applications of SAI shall aid grounding the term in tangible scenarios.

Consequently, "Spiritual AI" pertains to AI systems capable of identifying and deciphering human consciousness from text via natural language processing, audio through speech analysis, video by examining facial movements, and other multimodal data sources like gait analysis and physiological signals. While experts in AI and neuroscience concur that present AI models lack true consciousness or mindfulness, they can emulate life's vitality by showcasing empathy, perceptions, personality, and morality.

The idea of Spiritual AI—integrating principles of spirituality into artificial intelligence—seems a significant departure from our traditional understanding of AI. However, its need becomes more evident as we delve into the role of technology in our lives, specifically concerning mental health and well-being.

3.9 Concluding Remarks

We examine various aspects of spiritual intelligence, each contributing to a broader and richer understanding of what it means to live a spiritually aware life and spiritual wellness.

We began by exploring consciousness, which forms the foundation of spiritual intelligence. We discussed mindfulness, the ability to stay present and aware, and how transcending mere rationality can open up deeper levels of understanding and insight. Moving on to grace, we touched on the universal life force and how it interconnects all living beings. Understanding life force energy helps us appreciate the subtle and profound ways in which grace flows through our lives, offering support and guidance. Meaning suggests finding purpose and significance in our experiences can profoundly impact our spiritual well-being. Transcendence emphasizes the importance of going beyond everyday concerns to connect with higher aspects of ourselves and the universe, leading to the personal transformation. Understanding deeper truths about our nature and the world helps us navigate life with greater clarity and authenticity. Peaceful surrender to self leads to acceptance of our true nature with serenity, fostering inner peace and resilience. Inner directedness encourages us to focus on our inner values and intuition rather than external pressures, guiding us toward a more authentic and balanced life.

Embracing these dimensions can lead to deeper self-awareness and greater fulfillment. Spiritual AI highlights the use of advanced technology to enhance our spiritual wellness by tackling these dimensions of spirituality on the foundations of emotional quotient and intelligence quotient.

Chapter 4
Spiritual AI: Scope of Study

As we explore the concept of spiritual AI, we find that it intersects with many different fields, revealing the complex nature of spirituality. This integration offers deep insights, helping us to understand the spiritual phenomena more holistically and allowing AI to possibly enhance human spiritual experiences. Integrating spiritual AI with various fields opens up many possibilities. It addresses ethical issues, metaphysical questions, and the blending of technology with spiritual concepts. This interaction enriches our understanding of *existence* and *morality*, leading us to think deeply about AI's impact on beings' spiritual growth (Fig. 4.1).

4.1 Healthcare

Puchalski in 2002 underscores the necessity of integrating spiritual care into medical practice, suggesting that addressing the holistic needs of the patients, including their spiritual wellness, is crucial for providing truly compassionate care (Puchalski, 2001). The practical examples have profound impact on patients' psychological well-being and their experience of healthcare. Carl E. Thoresen from Stanford University examines the relationship between spirituality and health from a scientific perspective (Thoresen, 1999). Thoresen illustrates that while there is promising evidence suggesting a positive relationship between spirituality and health, many studies lack comprehensive assessments and adequate controls. He emphasizes the necessity of methodologically diverse and culturally sensitive research to study the impact of spirituality on various health outcomes. Research in these areas may reflect the societal shifts, evolving perceptions, and the quest for deeper connectedness in fragmented social landscape.

In an era where AI is transforming healthcare, the integration of SI with AI addresses multiple dimensions of human intelligence that are efficient, empathetic, and culturally sensitive. For instance, AI chatbots (Chen et al., 2023) incorporating

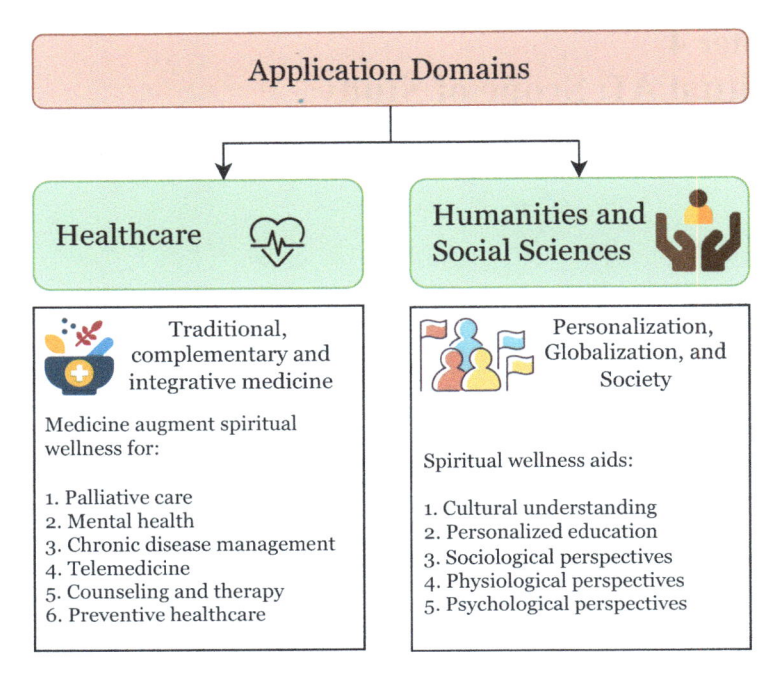

Fig. 4.1 Applications of Spiritual AI

SI to detect emotional states, making healthcare AI more empathetic, ethical, and culturally sensitive. By considering diverse socioeconomic backgrounds, Spiritual AI promotes a more holistic and equitable approach to healthcare delivery. Facilitating heightened empathy and sensitivity, emotional intelligence elevates the therapeutic ambience in mental healthcare settings, ensuring more comforting experience for participants. SAI enhances holistic well-being, and by focusing on the collective welfare, it transcends individual health, emphasizing communal harmony and shared well-being. To embrace the spiritual element, CAM has recently gained importance.

Alternative Medicine is a broad set of healthcare practices that are not part of a country's traditional or conventional medical system (often referred to as Western medicine). These practices are based on historical or cultural traditions, rather than on scientific evidence.

Holistic medicine considers people and how they interact with their environment. It emphasizes the connection of mind, body, and spirit to achieve optimal health and well-being by uncovering the underlying causes of diseases and disorders, rather than merely treating or managing the symptoms.

As we further observe different alternative medicines and holistic medicine, we discuss them in Tables 4.1 and 4.2, respectively.

4.1 Healthcare

Table 4.1 Alternative medicine and their description

Alternative medicine	Description
Acupuncture	Traditional Chinese technique of inserting thin needles at specific points
Homeopathy	Treatment involving small doses of substances that would produce symptoms of the illness in larger amounts
Naturopathy	Treatment emphasizing natural remedies like diet and exercise
Ayurveda	Ancient Indian medicinal system
Chiropractic	Manipulation of the spine and other body structures
Traditional Chinese Medicine (TCM)	Ancient Chinese medicine using various methods like acupuncture, cupping, and herbal remedies
Biofeedback	Technique for learning to control bodily functions using monitoring devices
Reiki	Japanese energy healing technique
Osteopathy	Focus on the body's musculoskeletal system
Anthroposophical medicine	Integrates spiritual and physiological processes in diagnosis and treatment
Herbal medicine	Use of plants or plant extracts for medicinal purposes
Bach Flower Remedies	Dilutions of flower material developed by Edward Bach
Iridology	Diagnosing by examining the iris
Cupping	Using heated cups to create suction on the skin
Moxibustion	Burning dried mugwort on or near the skin
Colon hydrotherapy	Flushing the colon with water
Aromatherapy	Using essential oils for healing
Shiatsu	Japanese bodywork based on TCM concepts
Reflexology	Applying pressure to reflex zones on feet, hands, or ears

4.1.1 Palliative Care

Palliative care focuses on providing relief to individuals suffering from serious illnesses, aiming to alleviate symptoms, pain, and stress associated with the condition. The essence of palliative care is not to cure but to improve the quality of life for both patients and their families.

> Shruti, a patient diagnosed with advanced cancer may receive palliative care to manage pain and discomfort from the disease and its treatment, coupled with emotional and spiritual support to navigate the psychological and existential challenges posed by the serious illness. The family of the patient would also receive support and guidance to cope with the challenges and stresses associated with their loved one's illness.

Palliative care is integrally associated with Cognitive Intelligence (CI), Emotional Intelligence (EI), and Social Intelligence (SoI), as it emphasizes a holistic approach to patient care. CI aids in the decision-making process regarding treatment options, symptom management, and care planning. Patients and families benefit from EI by managing their emotional responses to illness, maintaining emotional well-being, and developing resilience. Healthcare providers use SoI to collaborate

Table 4.2 Holistic medicine and their description

Holistic medicine	Description
Mind–body techniques	Includes meditation, visualization, and yoga
Nutritional therapy	Diet-based healing
Energy healing	Such as qi gong, chakra balancing, therapeutic touch
Herbalism	Comprehensive use of plant remedies
Aromatherapy	Therapeutic use of aromatic plant extracts and essential oils
Crystal healing	Using crystals and gemstones for therapeutic purposes
Sound therapy	Use of sound waves and harmonics for healing
Spiritual healing	Tapping into spiritual or divine forces for healing
Color therapy	Using color to influence health and well-being
Detox therapy	Various methods used to cleanse the body
Hydrotherapy	Using water for pain relief and treatment
Magnetic therapy	Using magnets to treat ailments
Bio-resonance therapy	Using electromagnetic waves to diagnose and treat
Balneotherapy	Therapeutic bathing using mineral-infused water
Tai Chi	Martial art known for its health benefits
Hypnotherapy	Clinical use of hypnosis
Guided Imagery	Mental visualization to improve mood and physical well-being
Qi Gong	Chinese practice integrating physical postures, breathing, and focused intention
Rolfing	Form of deep tissue massage

with other professionals, communicate effectively with patients and their families, and facilitate social support.

The impending loss of a loved one often triggers profound existential questions and concerns, requiring holistic care that addresses physical, emotional, social, and spiritual needs. In this context, existential distress is a significant concern, intensifying the need for end-of-life care that genuinely integrates every aspect of an individual's being. The use of SAI ensures that people facing life-limiting illnesses receive care that respects and understands their multifaceted needs.

Spiritual AI has leveraged computational techniques to understand and interpret patients' spiritual beliefs, experiences, and preferences, allowing the provision of spiritual care for spiritual wellness:

- There are several AI-powered chatbots and virtual health assistants (VHA), like Woebot (Prochaska et al., 2021) and Wysa (Inkster et al., 2018), which provide personalized feedback using cognitive-behavioral therapy principles.
- MyDirectives (formerly ADVault) is an example of a platform that allows individuals to share their end-of-life care preferences (Gazarian et al., 2019). MyDirectives helps people record their medical treatment wishes, palliative and hospice care preferences, organ donation status, and other critical information on the device, and in the format, that is most convenient for them.

- Wearables like Fitbit and Apple Watch can potentially integrate SAI to monitor spiritual wellness, through analyzing data related to stress, meditation, and other related metrics (Akdevelioglu et al., 2022).

AI is increasingly being used in hospice care for predictive analytics, identifying patients who may be in need of hospice care. Hospitals and healthcare providers like Mayo Clinic and Johns Hopkins are incorporating holistic approaches in palliative care, including addressing spiritual concerns. The development of SAI in palliative care holds the potential to further revolutionize personalized patient care.

4.1.2 Mental Health

Mental health is comprised of cognitive, emotional, and social intelligence. Headspace (O'Daffer et al., 2022), an app focusing on meditation and mindfulness, uses data to offer personalized interventions aligned with one's beliefs. The SAI-powered app suggests interventions like mindfulness exercises or meditative practices, congruent with user's spiritual beliefs. Industry leaders like BetterHelp[1] and TalkSpace[2] may want to enhance their platforms with SAI to enrich their therapeutic models by integrating spiritual insights into their counseling services.

Woebot uses the principles of Cognitive-Behavioral Therapy (CBT) and demonstrates potential for SAI integration. Similarly, 7 Cups,[3] an online platform, combines elements of emotional and social intelligence to connect individuals with trained listeners and holds the potential for integrating SAI. Similarly, Calm,[4] Insights timer,[5] and virtual reality driven VRHealth[6] have the potential to enrich its experience spiritually.

Leading mental hospitals in the USA offer treatment with ultra-modern facilities such as advanced neuroimaging techniques, electroconvulsive therapy, and transcranial magnetic stimulation, esketamine for treatment-resistant depression, and precision medicine for personalized treatment. Despite the high cost of treatment, patients seek treatment in the USA due to their high standards and the best hospitals for psychiatry such as McLean Hospital, John Hoskins Medicine, and Mayo Clinic. According to 2010 Complementary and Alternative Medicine Survey of Hospitals,[7] 42% of the 714 hospitals across the country have responded to patients' demands and integrate CAM services with conventional, routine services.

---

[1] https://www.betterhelp.com/

[2] https://www.talkspace.com/

[3] https://www.7cups.com/

[4] https://www.calm.com/

[5] https://insighttimer.com/

[6] https://vrhealth.institute/

[7] https://www.itcim.org/a-cam-survey-in-us-hospitals

AI technologies can complement and enhance TCIM, aligning with the shared objectives of researchers from both fields in providing personalized treatment plans, predicting health trends, and enhancing patient engagement (Ng et al., 2024). Innovations like Replika (Possati, 2023) engage users in text-based conversations using cognitive and emotional intelligence, potentially enhancing its adaption to user's spiritual beliefs. Amidst the growing interest in developing task-autonomous AI for automated mental healthcare using *generative AI models*, the research community has recently outlined the ethical requirements and defined beneficial default behaviors for AI agents in the context of mental health support (Grabb et al., 2024; Garg et al., 2024b).

4.1.3 Chronic Disease Management

Chronic Disease Management is designed to deal with long-term illnesses, emphasizing the control of the chronic health conditions and prevention of further deterioration through CI, EI, and SoI. CI understands and processes medical information, manages symptoms, and adheres to treatment plans. EI manages the psychological stress associated with chronic conditions, and SoI is essential for interacting with healthcare providers, caregivers, and support groups.

Real-world applications of Spiritual AI in chronic disease management are MySugr for diabetes management (Debong et al., 2019), and wearable devices such as Apple watch, Xiaomi Mi Band, Biostrap, Huawei Band, Garmin Vivosmart, Whoop Strap, and Samsung Galaxy Fit that are used for monitoring and managing chronic conditions through fitness tracking features. The advanced multisport watches (Polar Vantage V2) (Gruber et al., 2022) and hybrid smartwatches like (Withings Steel HR Sport) (Connelly et al., 2021) have health tracking features. The Oura Ring[8] provides sleep, activity, and readiness tracking, and the Suunto 9 Baro is designed for long training sessions.

Specialized wearable like the Wahoo Tickr X is a chest-worn heart rate monitor known for its precision, and the Amazfit GTR 3 Pro integrates smart features with varied sports modes. These innovations represent the extensive diversity and capability, offering solutions for fitness tracking and health monitoring.

4.1.4 Telemedicine and Remote Patient Monitoring (RPM)

The Community Preventive Services Task Force recommends TeleHealth interventions for reducing chronic disease risk factors in patients and managing chronic

---

[8] https://ouraring.com/

4.1 Healthcare

diseases.[9] By leveraging digital communication technologies, telemedicine ensures that patients can access medical consultations, online, bridging geographical barriers, and delivering quality care to patients irrespective of their location. In telemedicine, three interesting dimensions are *technical know-how, empathetic patient interactions, and effective digital collaboration*:

1. **Technical expertise** ensures that medical professionals can use telecommunication tools with precision, facilitating seamless virtual consultations.
2. **Empathy remains at the forefront of patient interactions**, as understanding and addressing the concerns of patients from a distance can pose challenges.
3. **Effective digital collaboration** is integral for a team of healthcare providers to work in tandem, ensuring that a patient's care plan is coherent and comprehensive.

The potential of Spiritual AI in telemedicine is observed through platforms like Teladoc or Doctor on Demand (Jain et al., 2019). Infusing Spiritual AI into these platforms enables them to comprehend the consciousness and mindfulness of patients. As telemedicine grows, so does the scope for wearables and RPM devices. Innovations such as the Philips Health Watch (OShea et al., 2024) and the Withings BPM Connect are revolutionizing RPM by offering insights for heart rate variability, blood pressure, and other vital metrics. The Omron HeartGuide, a wristwatch, goes a step further by offering *oscillometric blood pressure measurements* typically found in arm cuffs.

For more specialized needs, the *Eko CORE digital stethoscope* is transforming cardiac patient care by amplifying, recording, and analyzing heart sounds. Medtronic Care Management Services offer data collection from patients through patient engagement, and the Fitbit Care platform has daily activity tracking with health coaching. Such diverse offerings exemplify the current state of telemedicine and RPM devices—ranging from fundamental health insights to specialized medical utilities. Integrating Spiritual AI into their telemedicine protocols can allow for a more well-rounded approach to patient care, integrating spiritual guidance alongside clinical recommendations.

4.1.5 Counseling and Therapy

Counseling and therapy are therapeutic techniques aimed at improving mental health and addressing psychological concerns. SAI offers a deeper, quantified understanding of individuals' spiritual wellness. By integrating SAI, therapy platforms can analyze spiritual narratives, guiding therapists in customizing interventions. For example, online therapy platforms like BetterHelp could incorporate SAI to help users navigate spiritual dilemmas. The convergence of meditation apps

---

[9] https://www.thecommunityguide.org/

Table 4.3 A list of meditation Apps

App	Description
Headspace	Offers guided meditations and techniques for stress relief
Simple habit	Experts-driven unique meditation platform
Meditation studio	Features a mobile-first approach to guided relaxation
Being	Delivers personalized spiritual wisdom
Balance–Meditation & Sleep	Tailors daily meditation experiences based on user feedback
Calmind	Combines neuroscience and psychoacoustics for mental fitness
5 Minute Escapes	Provides quick short, guided relaxation sessions
Stop Breathe & Think	Promotes mindfulness and compassion
Mindfulness Calculator by Mist	Measures users' levels of mindfulness
enso	Sending reminders to breathe and relax
Breathe Meditations for Mac	Helps manage stress and improve focus, especially at work

like Calm with SAI presents opportunities for customized spiritual exercises. Healthcare institutions, such as the Mayo Clinic, foresee research with technology and spirituality, emphasizing a holistic approach to mental health.[10]

The world of mental health is undergoing transformation with the rise of generative AI. AI can process datasets, offering therapists insights into patients' behaviors. Platforms like Replika adapt to users' emotions over time through meaningful conversations. VMAs, like Moodfit and Wysa, track emotional health and provide coping strategies, keeping therapists informed between sessions. Merging AI with spiritual practices such as Aurahealth[11] provides mindfulness meditations customized to users' emotions. A list of other meditation Apps is given in Table 4.3.

Despite its potential, there are hurdles. Training data determines AI's effectiveness. For instance, Microsoft's Tay became problematic due to biased data, emphasizing the need for reliable data sources. With an increasing number of female therapists, tools like Ellie, designed for PTSD patients, must be aware of gender nuances. Data privacy is paramount, necessitating robust protection protocols. While AI's role is expanding, the significance of human interaction remains paramount.

4.1.6 Preventive Healthcare

Preventive healthcare or preventive care takes measures to prevent diseases or injuries rather than treating them once they have occurred. The primary goal is to protect, promote, and maintain health and well-being and to prevent disease,

---

[10] https://www.mayoclinicplatform.org/2022/06/23/at-the-intersection-of-technology-and-spirituality/

[11] https://www.aurahealth.io/

disability, and death. Preventive healthcare is categorized into primary, secondary, and tertiary prevention. *Primary Prevention* aims to prevent disease or injury before it occurs by preventing exposures to hazards that cause disease or injury. It alters unhealthy or unsafe behaviors and increases resistance to disease. *Secondary Prevention* aims to reduce the impact of a disease or injury that has already occurred by detecting and treating disease as soon as possible to halt or slow its progress. It encourages people to their original health and function. *Tertiary Prevention* aims to soften the impact of ongoing illness or injury that has lasting effects by managing long-term, complex health problems, improving the quality of life, and life expectancy.

Toward SAI for Healthcare
Health is not just the absence of physical illness but encompasses emotional, mental, and spiritual well-being. SAI can analyze an individual's spiritual wellness. SAI tools may offer personalized meditation, prayer, or mindfulness exercises. Regular engagement with these practices can reduce stress, a major risk factor for numerous diseases, and enhance emotional regulation. SAI can provide recommendations for lifestyle changes that align with their values and convictions. SAI can be used to create educational modules or resources tailored to an individual's spiritual beliefs, offering guidance on healthy practices, dietary habits, and lifestyle choices that align with their spiritual wellness.

While the potential of SAI in preventive healthcare is vast, it is essential to approach its integration with sensitivity, respect for individual beliefs, and strict data privacy standards. The spiritual dimension is deeply personal, and its incorporation into healthcare, especially with AI, should be done thoughtfully and ethically.

In the era marked by the feminization of aging, there lies an unprecedented opportunity to intertwine geriatric care with SAI. Envision a future where AI-driven platforms, tailored to the spiritual nuances of older women, offer personalized therapeutic interventions, facilitating mental well-being, community connection, and informed end-of-life decisions. As we pen the next sections of healthcare innovation, it is important to learn the fusion of spirituality and AI in geriatric care, suggesting solutions that resonate with both mind and soul of our aging population.

4.2 Humanities and Social Sciences

Infusing SAI with Humanities and Social Sciences offers a transformative approach to understand human experiences and societal dynamics. This synergy can lead to a deeper cultural understanding, personalized educational content, and more empathetic social interactions. It facilitates a holistic view of societal trends, aids in preserving cultural traditions, promotes universally acceptable ethical frameworks, and supports personalized therapeutic interventions. Such integration can also enhance literary and art analysis, inspire interdisciplinary research, and strengthen community bonds, fostering a more interconnected and harmonious world.

4.2.1 Cultural Understanding

The unparalleled insight into cultural intricacies can be bolstered by Spiritual AI, with its ability to process and understand vast the cultural information, uncovering the human values, beliefs, and practices from varied civilizations. It holds the potential to foster mutual respect and understanding in multicultural societies, thus diminishing prejudices and biases.

Cultures are reflections of collective human experiences, and one can understand culture by delving into its spiritual and literary texts. It is often passed down through generations, and given the sheer volume and complexity of cultural texts available from around the world, gaining meaningful insights can be overwhelming. SAI may assist in analyzing and comparing these texts at deeper levels that might be challenging for the human mind alone. By ML and NLP capabilities, Spiritual AI can dissect, interpret, and present the core spiritual themes of various texts, highlighting the essence of the culture they stem from.

Exercise Begin by choosing three texts from diverse cultures. For instance, you might select the "Bhagavad Gita" from Indian culture, the "Tao Te Ching" from Chinese traditions, and "Rumi's Poetry" from Persian Sufism. Input these texts into the *Spiritual AI tool designed for cultural analysis* or assume that you did this in lieu of its absence. Ensure you are using a platform that can highlight key themes, motifs, and underlying values. Once AI provides its insights, study the themes and interpretations it presents. For the "Bhagavad Gita," it might highlight the concepts of duty, righteousness, and the nature of the universe. For the "Tao Te Ching," themes might revolve around balance, simplicity, and the ineffable nature of Tao. "Rumi's Poetry" might illuminate love, the quest for the divine, and the transcendence of the self.

> Begin your essay by introducing the texts and their cultural backgrounds. Proceed to detail the insights the Spiritual AI tool provided for each text. While elucidating the themes, delve into their significance in the respective cultures. The core of your essay should focus on drawing comparisons and contrasts. Despite the apparent diversity, you might find universal human experiences and values echoed across these texts. For instance, all three texts might touch upon the idea of a higher purpose or the pursuit of inner peace, albeit in different terminologies and contexts. Conclude your essay with reflections on the power of literature and spirituality in bridging cultural gaps. Highlight how, despite our diverse expressions, the essence of human experience and aspiration remains strikingly similar across cultures.

Imagine how interesting would it be to gain useful insights with this **enthnocomputing**. Broadly, ethnocomputing is the study of the cultural relevance and context of computing, computational practices, and computing artifacts within various cultural and societal groups. It emphasizes understanding and acknowledging that computing, and how people interact with computational systems, is deeply embedded within cultural practices, values, and norms. Ethnocomputing challenges the one-size-fits-all approach to computing by recognizing that different cultures may have different understandings, applications, and interactions with technology. It is an interdisciplinary approach, combining insights from anthropology, cultural studies, and computer science to explore the diverse ways in which cultures intersect with computing.

For example, the way one community or culture approaches problem-solving, algorithmic thinking, or even user interface design might differ from another due to cultural norms, historical contexts, and societal values. Recognizing and studying these variations helps in creating more inclusive technologies and broadening the understanding of computing in global contexts. Thus, infusing ethnocomputing through spiritual AI can lead to a harmonious merger of cultural understanding and spiritual sensitivity in computational systems:

- **Cultural–Spiritual Curriculum Design**: Educational platforms can provide content based on both cultural and spiritual backgrounds. For instance, lessons on environmental science might be grounded in indigenous spiritual tales that emphasize harmony with nature.
- **Personalized Meditation and Wellness Platforms**: Such platforms could adapt meditation techniques, rituals, or exercises based on both the spiritual practices and cultural traditions of users, offering a unique blend that resonates on a personal level. For instance, various wellness platforms aim to enhance employee health and engagement. *Wellics* provides insights into personal well-being to promote healthier habits, while *MoveSpring* leverages physical challenges to connect remote teams. *Metta*, catering to the Latin American market, emphasizes habit formation and team building. *Wellable* delivers a global wellness solution with diverse resources, and *Woliba* focuses on community engagement and wellness education. *Vantage* Fit uses activity tracking to inspire a healthier lifestyle, and *BurnAlong* offers a broad spectrum of health classes for multiple fitness levels.
- **Cultural Heritage Preservation**: Many indigenous languages, rituals, and spiritual customs are at risk of fading away. Ethnocomputing-driven systems, sensitive to spiritual nuances, can digitally document, interpret, and showcase these practices, allowing younger generations and global audiences to access and appreciate them.

Spiritual AI can play a pivotal role in preserving cultural heritage by digitally archiving and restoring significant artifacts, offering immersive spiritual storytelling through AR and VR, aiding in the preservation of ancient languages, and setting ethical guidelines for the treatment of sacred items. It can foster interfaith dialogues by highlighting shared spiritual narratives, predict potential threats to cultural sites using data analytics, and enhance museum experiences by providing visitors with enriched insights into the spiritual essence of displayed artifacts. This fusion of technology and spirituality offers a holistic approach to safeguarding the rich cultural diversity of the world.

4.2.2 Personalized Learning in Education

The realm of education stands to benefit greatly from SAI. By gaining an understanding of an individual's spiritual and cultural backdrop, educational platforms can tailor content in the fields of humanities and social sciences. This personalization ensures that the content resonates more deeply with students, offering them a perspective that aligns with their inherent beliefs and values, making learning more relatable and impactful.

History is not merely a linear recounting of past events; it is deeply intertwined with the cultural, spiritual, and sociopolitical contexts of various epochs. Imagine a high-school student, Dhwani, with Buddhist lineage embarking on a course about World Religions. Instead of a generalized curriculum, Dhwani would benefit from starting the course with an in-depth exploration of Siddhartha Gautama's journey to enlightenment, the foundational principles of the Eightfold Path, and the spread of Buddhism across Asia. As the course progresses, this foundation in Buddhism can serve as a reference point, drawing parallels and contrasts with other world religions. This tailored approach not only fosters a deeper connection to the material but also facilitates a broader understanding of global religious dynamics through a familiar lens.

Literature is a reflection of the human experience that resonates with our spiritual and emotional selves. SAI can understand student's innermost inclinations and recommending readings that echo those sentiments for spiritual wellness. Take, for example, a student who feels a profound spiritual connection with nature. Instead of a generic reading list, the AI might suggest works like John Muir Thoreau's writings (Cushman, 2023), which encapsulate profound reflections on nature and its spiritual significance. Such personalized reading not only makes literature more relatable but also fosters a deeper appreciation of the universality of certain themes across different writings.

Philosophy deals with the most intense questions of existence, morality, and purpose. By presenting ethical dilemmas in a context that aligns with the student's beliefs, the exploration becomes deeply personal, leading to a more enriched and nuanced understanding of philosophical ethics.

Integrating Spiritual AI into education is about recognizing and valuing the diverse spiritual tapestries that students bring to the classroom. Designing a Spiritual AI model that understands different cultural values would be an interesting innovation in the near future. The model would use AI to spot and fix its own biases. Some models will create online spaces like digital museums or games where people can explore ethical problems from various cultures. Making school lessons more personal by understanding a student's cultural background can teach about inequalities and help solve disagreements between different cultural groups. Transform counseling for immigrant students may help in adapting new environments.

4.2.3 Sociological Perspectives

Sociological perspectives offer invaluable insights, connecting individual consciousness to the broader societal fabric. Central to our exploration are Emile Durkheim's concept of **collective consciousness**. For example:

- **National Anthems and Flags**: These symbols invoke a sense of unity and shared identity among citizens of a nation. Singing the anthem or saluting the flag becomes a collective act, representing shared ideals and values.
- **Cultural Festivals**: Events like India's Diwali or the U.S.'s Fourth of July celebrate shared histories, myths, and values, promoting a sense of collective belonging.

Durkheim's idea of collective consciousness connects beautifully with the spiritual notion of the Universal Life Force—a ubiquitous energy or vitality present everywhere and in every being.

At the heart of Durkheim's collective consciousness is the idea of interconnectedness—how shared beliefs and values create a cohesive societal fabric. This mirrors the essence of the *Universal Life Force*, which is all about interconnectedness—how every being and every particle in the universe is linked through a common energy. The strength of collective consciousness can be so powerful that it transcends individual beliefs and becomes a societal truth. Similarly, the Universal Life Force, when channeled or harnessed, can bring about profound spiritual awakenings, healing, and transformations.

> For example, think of collective meditation or prayer events, where thousands gather with a shared intention, whether it's world peace, healing, or gratitude.

The collective energy generated in such gatherings is palpable and can be seen as a confluence of collective consciousness and the Universal Life Force.

Social Constructivism Social Constructivism posits that our understanding of reality is constructed through societal interactions and shared beliefs (Vygotsky & Cole, 2018). The collective interpretations of a society shape individual perceptions. Societal norms and values become "realities" that individuals abide by.

Symbolic Interactionism Developed by George Herbert Mead and Charles Horton Cooley, **Symbolic Interactionism** suggests that people act based on the meaning they ascribe to things, and these meanings arise from social interaction (Jacobs, 2009). Collective events, like meditation or prayer sessions, become meaningful symbols of unity and shared purpose. As individuals interact and share these experiences, the symbolic meaning of event can amplify, resonating with both the collective consciousness and the Universal Life Force.

Cosmopolitanism Proposed by Ulrich Beck and others, **cosmopolitanism** suggests that in a world risk society, individuals and groups need to think beyond local and national interests and embrace a global consciousness (Beck, 2011). The Universal Life Force concept can be seen as a spiritual reflection of cosmopolitan thinking, recognizing a shared life force or energy promotes global unity and interconnectedness.

4.2.4 Psychological Perspectives

Spiritual beliefs and practices offer coping mechanisms in times of distress. It provides solace, offers hope, and can be a source of internal strength. Studies have shown that people who engage in regular spiritual practices, or who have strong spiritual or religious beliefs, tend to have better mental health, lower levels of depression and anxiety, and a greater sense of overall well-being (Plante & Sharma, 2001). Meditation, a spiritual practice, enhances emotional regulation, improves focus and concentration, reduces stress levels, and even brings about structural changes in the brain, particularly in areas related to attention and sensory processing (Afonso et al., 2020).

Consciousness is the moment-to-moment awareness of our internal thoughts and feelings, as well as our external environment. It is the continuous narrative our mind constructs, which lets us experience and interpret the world around us. Psychology has always been intrigued by altered states of consciousness, achieved through meditation, hypnosis, or even drug use. Dreams, being a natural altered state that everyone experiences, have been subjects of fascination, from Sigmund Freud's psychoanalytic interpretations to modern neurological explorations (Freud, 1989).

Bernard Baars' Global Workspace Theory (GWT) provides an interesting metaphor for consciousness (Baars, 2005). Imagine the mind as a vast office space with many workers (unconscious processes) busy with their tasks. Every so often, one task (or process) becomes the center of attention, spotlighted in a central workspace where it gains prominence and can be worked upon, analyzed, or acted upon.

Neural Network Theories in cognitive psychology emphasize the interconnected systems of neurons within the brain and their crucial role in dictating how we think, feel, remember, and perceive the world. These networks form the backbone of our cognitive processes and offer a bridge between the observable physical structures of the brain and the intangible phenomena of mind and consciousness.

Donald Hebb, a pioneering neuropsychologist, proposed that "neurons that fire together, wire together" (Munakata & Pfaffly, 2004). This principle, often termed *Hebbian learning*, implies that when two neurons are activated simultaneously, the connection between them strengthens. Conversely, if they do not synchronize their firing, the connection weakens. This simple rule forms the basis for how neural networks might evolve and adapt over time.

Our neural connections are not static; they dynamically adapt based on our experiences. As we encounter new information or practice a skill, the relevant neural networks reshape themselves to encode this experience. This plasticity ensures that our brains remain adaptable throughout our lives.

Distributed Representation Unlike a computer, where a specific bit of information is stored at a specific location, the brain uses distributed representation. A single memory or thought is likely represented by the activity across a vast network of neurons. This distributed nature provides robustness to our memories and cognitive processes. Even if some neurons are damaged, the network can often still function due to this distributed coding.

Pattern Recognition and Generalization Neural networks in the brain excel at recognizing patterns. When faced with a familiar situation or stimulus, the brain activates the corresponding neural pattern. Moreover, due to the interconnected nature of these networks, the brain can generalize from past experiences, allowing us to adapt to novel yet similar situations.

While the specific neural basis of consciousness remains one of the greatest mysteries in neuroscience, some theories such as GWT suggest that the synchronous firing of large-scale neural networks might underlie conscious awareness. This means that our conscious experiences could emerge from the coordinated activity across vast and interconnected networks of neurons.

4.2.5 Philosophical Perspectives and Metaphysics

At its core, spirituality is often a philosophical journey. It probes questions like:
> Why are we here?
> Is there a higher purpose?
> What is the ultimate nature of reality?

These inquiries are deeply metaphysical. As science pushes the boundaries of understanding and within the realms of quantum mechanics, certain phenomena defy classical understanding, leading to interpretations that resonate with spiritual and metaphysical perspectives. This confluence is particularly evident when discussing topics like *consciousness* and *the nature of reality*.

Metaphysics is a branch of philosophy that examines the fundamental nature of reality, including the relationship between matter and mind, substance and attribute, and potentiality and actuality. It seeks answers to questions that often elude empirical investigation. Central to metaphysical inquiry is the origin of consciousness.

> How does subjective experience arise?
> Is consciousness a fundamental aspect of the universe, akin to space and time, or is it an emergent property of complex systems such as the brain?

While the parallels between quantum mechanics and spiritual notions are intriguing, many physicists and philosophers caution against overextending quantum theories into the realms of metaphysics and spirituality without rigorous scientific backing. While quantum mechanics has empirical support at the microscopic scale, its implications for macroscopic phenomena, especially something as complex as consciousness, remain speculative and are a topic of ongoing debate.

Quantum mechanics, with its wave-particle duality, superposition, and entanglement, presents a view of reality that challenges our classical intuitions. Some interpretations of these phenomena resonate with mystical and spiritual perspectives, leading to the emergence of *quantum mysticism* (Clayton, 2004). One of the most mystifying findings in quantum mechanics is *entanglement*, where particles become interconnected such that the state of one instantly affects the state of the other, regardless of the distance between them. This phenomenon resonates with mystical notions of interconnectedness and universal oneness, a core tenet in many spiritual traditions.

The orchestrated objective reduction (Orch-OR) theory by Roger Penrose and Stuart Hameroff suggests that consciousness arises from quantum processes within the brain (Stuart, 1998). They proposed that quantum processes within the brain's microtubules might be responsible for consciousness.

While neuroscience largely operates within a materialistic framework, understanding the intricacies of mind can help in finding answers to metaphysical questions about mind and matter, free will, and the nature of self. Future discoveries in neuroscience might reveal more about the quantum processes of brain.

4.3 Concluding Remarks

As we wrap up our examination of the scope of study in Spiritual AI, it is clear that this field holds remarkable potential across a wide range of disciplines. Spiritual AI can impact healthcare, humanities, and social sciences, each in unique and meaningful ways. In healthcare, innovations can help improve quality of life, provide more accurate diagnoses, and make healthcare more accessible. From cultural understanding and personalized education to sociological and psychological insights, this technology can provide deeper insights and better experiences with spiritual wellness. The philosophical and metaphysical aspects of Spiritual AI invite us to explore new disciplines associated with our spiritual wellness.

Chapter 5
Neurotransmitters: Foundations of Cognition

In recent years, the advancements in AI have reignited discussions around the potential consciousness of AI systems. Despite differing expert opinions, there exist scientific theories rooted in human studies that can be applied to AI systems to determine their potential consciousness. The rise of AI systems, especially large language models (LLM) mimicking human conversation, brings about ethical and societal considerations, making it essential to understand AI's relationship with consciousness.

As neuroscience progresses, our understanding of consciousness—that enigmatic sense of awareness and self-experience—undergoes perpetual refinement. The relationship between specific brain activities and the emergence of conscious experience remains at the forefront of scientific investigation.

5.1 Basics of Neuroscience

At the heart of our nervous system are neurons. These are unique and highly specialized cells that act as the main players in the complex communication system of our brain. Neurons work by sending and receiving electrical and chemical signals, allowing different parts of our brain and body to communicate with each other. Neurons connect with each other at junctions called synapses, where they transmit signals through neurotransmitters. This process allows our brain to process information, form memories, and coordinate actions. Additionally, the brain is divided into different regions, each responsible for specific functions like movement, sensation, and thinking.

Supporting the neurons are other types of cells, called glial cells, which provide structural support, protection, and nourishment to the neurons. The entire nervous system is a vast network that extends throughout our body, controlling everything from our heartbeat to our ability to solve complex problems. Understanding the basics of how neurons and the nervous system work helps us appreciate the incredible complexity and capability of our brains.

5.1.1 Neurons

Each neuron has several parts that help it function. One of the primary parts is the **dendrites**, which you can think of like tree branches. These branches reach out and catch signals from other nearby neurons, a bit like catching a ball. At the center of the neuron is what we call the **soma** or cell body that holds the nucleus (kind of like the brain of the neuron) and ensures that the neuron stays healthy.

Axon is like a long tail or wire attached to the neuron. Its main job is to carry electrical messages, known as action potentials, away from the neuron. These messages travel down the axon until they reach its end, where they cause the release of special chemicals called **neurotransmitters**.

Now, imagine two people standing at a distance, trying to pass a message without talking. They would use something like hand signals or written notes. In the neuron world, the gap they are throwing these messages across is called the **synapse**.

The neurotransmitters are like those notes, moving from one neuron's axon to the next neuron's dendrites, allowing them to communicate. The message can change from an electrical one to a chemical one and then back to electrical!

Different neurons have different functions. **Sensory neurons** report our brain about things we touch, see, or hear. **Motor neurons** give commands to our muscles about the movement. **Interneurons** refine the messages and make sure everything makes sense before it is relayed.

Neurons are responsible for transmitting information. Each neuron can form thousands of synaptic connections to other neurons, enabling intricate networks of communication pathways, travelled by neurotransmitters. The strength of synaptic connections can change based on experience—a process known as synaptic plasticity. This changeability underlies learning and memory, where repeated experiences or patterns of activity can strengthen or weaken connections.

5.1.2 Glial Cells

Other than neurons—brain's messaging system—there is another set of cells, called **glial cells** or just **"glia"**, that act as the support crew for brain. While they do not directly pass messages like neurons, they play several crucial roles in maintaining the brain's health and function.

Among the glial cells, **astrocytes** stand out with their star-like shape. These cells help create the blood–brain barrier, a protective layer that stops harmful substances from entering the brain. They regulate the levels of certain chemicals and nutrients in the brain and play a part in repairing the brain after an injury.

Microglia act like the brain's security team. They are always on the lookout, monitoring the brain environment. When they spot harmful invaders like germs, or damaged neurons, they spring into action, getting rid of them to keep the brain safe.

Oligodendrocytes in the central nervous system and **Schwann cells** in the outer parts of our body produce myelin. Myelin is an insulation around neuron's "wires" (axons) and makes sure the messages travel quickly and efficiently.

Ependymal cells line certain chambers in the brain and produce cerebrospinal fluid that acts as a cushion for the brain, protecting it from injury.

5.1.3 Structure and Organization of the Nervous System

The nervous system coordinates and regulates the body's activities. Broadly, it is divided into two main components: the Central Nervous System (CNS) and the Peripheral Nervous System (PNS). Each has distinct structures and functions, but together they create a synergistic system that allows us to sense, think, act, and respond to our environment.

Central Nervous System At the core of the CNS is the brain. As the central hub of our body's operations, the brain observed the cognitive functions, playing a big role in how we think, perceive, and act. Brain has the following components:

1. *Cerebrum*: Dominating a large portion of the brain's structure, the cerebrum deals with higher intellectual capacities such as abstract thinking/reasoning, daily operations, and conscious muscular activities.
2. *Cerebellum*: Nestled at the posterior part of the brain, the cerebellum maintains the equilibrium and coordination by refining our motor actions. It ensures that we move gracefully, maintaining balance and precise posture.
3. *Brainstem*: Acting as a crucial connector, the brainstem bridges the brain to the spinal cord. Beyond its structure, it controls essential automatic functions needed for survival, like breathing and heartbeat.

The spinal cord extends from the brain and is the main pathway for sending information to and from the brain. It is protected by the bones of the spine and ensures messages are sent efficiently. While it mostly acts as a messenger, the spinal cord can also perform certain reflexes on its own without the help of brain.

Peripheral Nervous System Beyond the CNS lies the **PNS**, which extends the central system's reach, connecting it to the farthest extremities of the body. The PNS comprises all neural structures outside the brain and spinal cord, playing a pivotal role in bridging the CNS with the rest of the body.

Somatic System This subsystem grants us voluntary control over our actions such as turning our head or playing a musical instrument. It instigates movement and sensory nerves, which ferry information from sensory receptors back to the brain.

Autonomic System Operating mostly beneath our conscious awareness, the autonomic system regulates involuntary bodily functions. It bifurcates into *Sympathetic Nervous System* and *Parasympathetic Nervous System*.

5.2 Neurological Correlates of Consciousness

The specific neurobiological processes in the brain that correspond with states and experiences of conscious awareness are called "Neurological Correlates of Consciousness." It is the search for where, when, and how conscious experience arises within the neural machinery of brain.

> Consciousness is a multifaceted phenomenon, encompassing our awareness of both the external environment and our internal thoughts, emotions, and sensations.

Neuroscientifically, it is not just the presence of brain activity that results in consciousness, but rather specific patterns and interconnections of such activity across various brain regions like the prefrontal cortex (PFC), the posterior parietal cortex (PPC), and the thalamus. These regions control executive functions, sensory processing, and information integration.

The brain is a complex organ with multiple regions and structures, each responsible for specific functions. Cerebrum, as discussed before, is the largest part of the brain that is covered by the cerebral cortex. The cerebrum is further divided into four lobes (see Fig. 5.1).

5.2.1 Frontal Lobe

The PFC is located at the very front of the *frontal lobe* region of the brain and is primarily associated with higher-order cognitive functions such as decision-making,

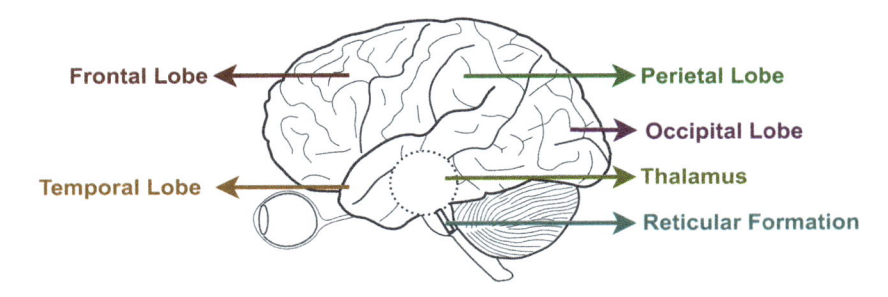

Fig. 5.1 Brain regions closely associated with various aspects of consciousness

planning, reasoning, and self-control. It is this region that helps us reflect upon ourselves, anticipate consequences of our actions, and make goal-oriented decisions.

The PFC can be likened to the "CEO" or the "conductor" of the brain. Located right behind the forehead, it is one of the last brain regions to fully develop, not reaching full maturity until around age 25. This lengthy development process underscores its importance and complexity. The PFC is not just a singular entity; it comprises different regions, each having its unique responsibilities. However, these regions work in tandem, much like the various departments in a company.

5.2.2 Parietal Lobe

The PPC is located in the upper and rear portions of the brain, straddling both hemispheres within the parietal lobe. Its position allows it to easily receive and integrate sensory data from various sources, making it an ideal hub for multisensory integration.

The PPC is pivotal for creating a "mental map" of our surroundings. It receives visual, auditory, and somatosensory inputs to produce a coherent spatial understanding. The PPC plays a significant role in this self-awareness. A disruption in this region can lead to syndromes where individuals might not recognize certain body parts as their own, illustrating the importance of the PPC in our conscious self-awareness. It integrates sensory feedback with motor commands, ensuring that our actions align with our conscious intentions. The PPC is deeply implicated in the processes of attention.

5.2.3 Thalamus

The thalamus is a paired, walnut-sized structure located deep within the brain, nestled between the cerebral cortex and the midbrain. Often termed the "gateway" or "relay station" to the cerebral cortex, the thalamus acts as a crucial intermediary that receives, processes, and forwards sensory information to specialized regions of the cortex.

With the exception of olfaction (sense of smell), virtually all sensory modalities—including vision, hearing, touch, and taste—send their primary inputs first to the thalamus. The thalamus plays a role in filtering and prioritizing the incoming information, ensuring that the most salient and relevant stimuli receive attention in the cortex. This role is crucial to prevent sensory overload. The thalamus's central position in the flow of sensory information and its dense connectivity with the cortex make it a prime candidate for involvement in consciousness. Some theories suggest that synchronized activity between the thalamus and certain cortical areas might underpin various conscious experiences.

5.2.4 Reticular Formation

The Reticular Activating System (RAS) is the central part of the brain that controls our states of consciousness. Located deep in the brainstem, this network of neurons is crucial for keeping us alert. It helps regulate our wakefulness, focus our attention, and manage the transitions between sleep and being awake.

Moreover, the dynamic nature of the RAS's operations is underscored by its close interplay with various neurotransmitters—chemical messengers in the brain. These neurotransmitters, such as acetylcholine, dopamine, and serotonin, among others, act as modulators. Depending on their concentrations and interactions, they can heighten our alertness, plunge us into deep slumber, or guide us through the intermediate stages of drowsiness and light sleep.

5.2.5 Occipital Lobe

The occipital lobe is responsible for visual processing and is located at the back of the brain. Given its primary role in visual perception, *occipital lobe* has been extensively studied in the context of consciousness, as vision is a major contributor to our conscious experience. Visual illusions offer a fascinating window into the workings of visual consciousness. They often reveal disparities between the physical properties of stimuli and our conscious perception of them. By studying how the occipital lobe processes these illusions, we can gain insights into the mechanisms of visual consciousness.

5.2.6 Temporal Lobe

The temporal lobe is one of the four main parts of the brain's outer layer. Located under the side groove of the brain on both sides, it is crucial for hearing, memory, emotions, and some aspects of vision. It is important for understanding consciousness because it helps with various processes that shape our conscious experience:

- **Auditory Processing**: The primary auditory cortex, located in the superior temporal gyrus, is crucial for processing auditory information. Conscious perception of sounds, from *binaural beats* (Ingendoh et al., 2023) to *singing bowls* (Seetharaman et al., 2024), largely stems from the activity in this region.
- **Memory**: The medial temporal structures, including the hippocampus and surrounding cortices, are essential for the formation and retrieval of declarative memories (memories of facts and events). Our conscious recollection of past events and our ability to mentally travel through time are intrinsically tied to these structures.

- **Visual Object Recognition**: Inferior temporal regions are involved in high-level visual processing and object recognition. This area helps us consciously recognize and categorize visual stimuli.
- **Emotion and Social Cognition**: The amygdala, a small almond-shaped structure deep within the temporal lobe, plays a pivotal role in emotion processing and social cognition. Conscious emotional experiences, especially those related to fear, are strongly linked to the amygdala.
- **Language Comprehension**: The posterior part of the superior temporal gyrus (in the dominant hemisphere) is known as Wernicke's area. This region is crucial for understanding language. Disruptions here can lead to a type of aphasia where speech is fluent but lacks meaning.

5.3 Impact of Neurotransmitters on Cognitive Abilities/Mental Health

Neurotransmitters are the chemical messengers of the nervous system. They play a pivotal role in transmitting signals across the synapses (gaps) between neurons. The release, reception, and reuptake of these neurotransmitters are critical to countless neural processes, including those that govern our thoughts, emotions, and behaviors:

- **Acetylcholine (ACh)**: ACh, PNS, and CNS, contributing significantly to attention, alertness, and mnemonic processes. Disturbingly, deficiencies in ACh have been observed in individuals suffering from Alzheimer's disease. As a result, numerous Alzheimer's treatments target either the augmentation of acetylcholine levels or the enhancement of its synaptic activity.
- **Glutamate**: Serving as the primary excitatory neurotransmitter in the brain, glutamate is responsible for neural signaling, learning, and memory processes. Excessive glutamate activity or release can instigate excitotoxicity, a process where neurons are damaged and killed. Aberrations in glutamate signaling have been implicated in conditions like schizophrenia and certain mood disorders.
- **Gamma-Aminobutyric Acid (GABA)**: GABA is the primary inhibitory neurotransmitter in the brain, acting as a neural pacifier. By inhibiting neural activity, GABA contributes to mood stabilization, reduction of neuronal excitability, and the prevention of overstimulation. A scarcity in GABAergic activity or disruptions in its signaling can cause mood-related disorders.
- **Dopamine**: Dopamine plays pivotal roles in systems governing reward, motivation, pleasure, and motor control. It also modulates mood. Parkinson's disease is characterized by a significant reduction in dopamine, specifically in regions vital for motor control. Conversely, overactivity of dopamine, especially in certain brain regions, is associated with schizophrenia. Dopamine's role in reward pathways also makes it central to the mechanics of addiction.
- **Serotonin (5-HT)**: Serotonin, also known as 5-HT, is a key neurotransmitter that mainly controls mood, appetite, and sleep. Often called the "feel-good" chemical,

serotonin helps maintain a positive mood and well-being. Low or imbalanced serotonin levels can lead to depression, anxiety, obsessive-compulsive behaviors, and sleep problems.
- **Norepinephrine (Noradrenaline)**: Norepinephrine, also known as noradrenaline, is a neurotransmitter pivotal for alertness, arousal, and orchestrating the body's stress responses. Disruptions in its levels or signaling can engender conditions such as depression, anxiety disorders, and attention-deficit hyperactivity disorder (ADHD).
- **Endorphins**: Endorphins are the body's innate painkillers, operating to dampen pain perception and elevate mood. They are often released in response to stress or physical activity, explaining the "runner's high" many experience post-exercise. Disruptions in endorphin levels or activity can influence both pain perception and mood regulation.
- **Anandamide**: Often referred to as the "bliss molecule," anandamide plays a pivotal role in feelings of pleasure and motivation. Beyond its euphoric attributes, it possesses antianxiety and antidepressant effects. Disruptions in its levels or activity can manifest as mood disturbances and altered pain perceptions.
- **Histamine**: While commonly associated with allergic responses, histamine also functions within the brain, playing roles in arousal, attention, and wakefulness. Elevated brain histamine levels have been associated with conditions like schizophrenia. Conversely, antihistamines, commonly used to treat allergic reactions, can block histamine activity, leading to side effects like drowsiness and cognitive dulling.
- **Oxytocin**: Oxytocin, sometimes colloquially labeled the "love hormone" or "social bonding hormone," has roles that extend from childbirth and breastfeeding to social bonding and trust. This neuropeptide bolsters social connections, deepening bonds between individuals. Disruptions or deficiencies in oxytocin signaling have been associated with social interaction deficits, as observed in conditions like autism spectrum disorders.

5.3.1 *Causes of Neurotransmitter Imbalances*

There are several reasons neurotransmitter levels might be disrupted, including:
- **Genetic Factors**: Genes can greatly affect neurotransmitter levels and how they work. Certain genetic changes can disrupt the normal production, release, and reabsorption of neurotransmitters. These genetic factors can make a person more likely to have neurotransmitter imbalances, impacting their mental and physical health.
- **Diet and Nutritional Intake**: The brain's ability to make neurotransmitters depends on the nutrients we eat. Lack of essential nutrients can slow down neurotransmitter production. For instance, the amino acid tryptophan is needed to make serotonin, which affects mood. Therefore, a diet missing tryptophan or other important nutrients can reduce the production of some neurotransmitters.

- **Chronic Stress**: Constant exposure to stressors, be they emotional, physical, or environmental, can have a profound impact on neurotransmitter reserves. Chronic stress can strain and exhaust the production of crucial neurotransmitters like serotonin and dopamine. This depletion can lead to mood disorders, fatigue, and other cognitive impairments.
- **Substance Abuse**: Drugs and alcohol can greatly change neurotransmitter levels in the brain. Drinking too much alcohol often can lower serotonin levels, causing mood problems. Drugs that increase neurotransmitter release can eventually deplete them, harming mental health.
- **Hormonal Fluctuations**: Neurotransmitters work together with other systems, especially the endocrine system. Changes in hormone levels during menstrual cycles, pregnancy, or menopause can affect neurotransmitter balance, causing mood swings, irritability, or depression.
- **Medicinal Impact**: Many psychiatric medications aim to fix neurotransmitter imbalances, but they can also cause them. By changing neurotransmitter levels, these medicines might create excesses or shortages, so regular monitoring and dosage adjustments are needed.
- **Environmental Exposures**: Our environment can affect our neurotransmitters. Long-term exposure to toxins, pollutants, or lack of sleep can harm neurons. This damage can disrupt neurotransmitter production and function, showing the importance of a healthy environment for brain health.

5.3.2 *Restoring Neurotransmitter Balance*

Neurotransmitters are the brain's chemical messengers, essential for mood, thinking, and mental health. Imbalances in neurotransmitters like serotonin and dopamine can lead to conditions like depression and anxiety. Fixing these imbalances is key to improving mental well-being.

Medication One of the main ways to fix neurotransmitter imbalances is with medication. Antidepressants like *selective serotonin reuptake inhibitors* (SSRIs) increase serotonin levels in the brain. Dopamine agonists are used for conditions like Parkinson's disease by boosting dopamine levels. These medications can be effective but may have side effects and need to be monitored by healthcare professionals.

The Role of Diet in Neurotransmitter Synthesis The brain needs various nutrients to make neurotransmitters. Amino acids, which are building blocks of proteins, help create neurotransmitters. For instance, tryptophan is needed for serotonin, and tyrosine is needed for dopamine. Omega-3 fatty acids, found in fish and flaxseeds, help keep neuron membranes flexible and can affect neurotransmitter function. Vitamins like B6, B12, and folate are also important for neurotransmitter production. A diet missing these nutrients can lead to imbalances.

Exercise and Neurotransmitter Levels Regular exercise is known to boost mood. It increases the release of endorphins, which are "feel-good" hormones that can reduce pain and create a sense of happiness. Exercise also raises levels of dopamine and serotonin, neurotransmitters that antidepressant medications target. This boost in neurotransmitters after exercise can lead to a better mood and improved well-being.

Light Therapy and Serotonin Production Bright light exposure, used in light therapy for conditions like seasonal affective disorder (SAD), can affect serotonin levels. Light therapy helps increase serotonin production, similar to natural daylight. This may explain why some people's moods change with the seasons. Light therapy shows how our environment can impact brain chemistry.

Cognitive-Behavioral Therapy (CBT) and Neurotransmitter Balance Cognitive-Behavioral Therapy (CBT) is a type of talk therapy that helps people recognize and change negative thinking patterns that cause emotional problems. By changing these thoughts, CBT can improve behavior and mood, which might also affect neurotransmitter systems. While the link between CBT and neurotransmitter levels is not fully clear, CBT is known to be an effective treatment for many mental health issues and can be used alone or with medication.

Avoiding Drugs and Alcohol to Preserve Neurotransmitter Health Drugs and alcohol can seriously disrupt neurotransmitter systems. For example, alcohol might boost neurotransmitter levels briefly, but long-term use can lower their natural production and effectiveness, leading to addiction and depression. Stopping these substances helps the brain recover and return to normal neurotransmitter levels, which is important for lasting mental health and stability.

Restoring neurotransmitter balance often requires a combined approach. This can include medication, diet changes, exercise, therapies like light therapy and CBT, and avoiding substance abuse. A personalized plan that includes these elements is usually the most effective way to address neurotransmitter imbalances and improve mental health.

5.4 Detecting Low Levels of Neurotransmitters

Measuring neurotransmitter levels, especially when they are low, is challenging and requires special methods. Neurotransmitters are important for regulating brain functions and consciousness. Here are some common techniques used to detect and measure these levels:

5.4.1 Microdialysis

Microdialysis is a technique used to measure chemicals in tissues, including the brain. A tiny probe is inserted into the brain and filled with a fluid, like saline. This fluid picks up small molecules, including neurotransmitters, from the surrounding brain tissue. The fluid, now containing these molecules, is collected and analyzed using methods like high-performance liquid chromatography (HPLC) to measure neurotransmitter levels.

Artificial intelligence (AI) can improve microdialysis by making it easier to detect and measure neurotransmitters. AI can turn chemical readings into digital data, improve its quality, and find patterns that might be missed otherwise. Using deep learning models, AI can detect small changes in neurotransmitter levels and provide more accurate results than traditional methods.

AI can enhance microdialysis by providing real-time adjustments, like changing saline flow rates, and creating easy-to-understand visuals of neurotransmitter data. It can also predict future changes in neurotransmitter levels, which is useful for research and medicine. As AI gathers more data, it improves its accuracy and adapts to new conditions.

However, using AI with microdialysis requires careful setup and might sometimes miss unique cases. Despite these challenges, combining AI with microdialysis can lead to more precise and predictive measurements of neurotransmitter levels. Microdialysis is a useful tool for real-time insights into neurotransmitter activity, but it is invasive and can potentially damage tissue. With proper training and equipment, these risks can be reduced, but they are still part of the procedure.

5.4.2 Cerebrospinal Fluid (CSF) Analysis

Cerebrospinal Fluid (CSF) Analysis is a key diagnostic method used to examine the chemical environment of the central nervous system. It involves a procedure called a *lumbar puncture or spinal tap*, where a needle is inserted into the lower spine to collect a sample of fluid surrounding the brain and spinal cord. This fluid is then analyzed in a lab for various substances, including neurotransmitter metabolites. However, the results may not always reflect the real-time activity of neurotransmitters in the brain.

CSF analysis is helpful for diagnosing and monitoring neurological disorders like multiple sclerosis, meningitis, and neurodegenerative diseases. It is also used in research to study psychiatric disorders like schizophrenia and depression. Since CSF is closely connected to the brain, its composition can show changes or problems in the brain. However, the procedure is more invasive than blood tests and can carry risks like headaches, back pain, or infections. Understanding the results can be complex and requires specialized knowledge.

Enhanced Diagnostic Accuracy AI can also predict the likelihood of developing certain conditions by analyzing past CSF data and specific biomarkers. By combining CSF data with other medical information (like genetic, imaging, or blood test results), AI can provide a complete picture of a patient's health, improving understanding and management of their conditions. AI can link CSF data with large medical databases to find new connections between neurotransmitter levels and various health issues.

Personalized Medicine Using AI to analyze CSF data can help customize treatments for each patient. For example, AI might suggest the best medications based on the patient's unique neurochemical profile. It can also monitor changes in CSF over time to track disease progression and treatment effectiveness.

Research and Development In drug research, AI can speed up the development of new medicines by showing how different compounds affect neurotransmitter levels. This helps researchers design drugs that target specific brain pathways more effectively. AI also helps understand complex brain interactions reflected in CSF, which is crucial for studying neurological and psychiatric disorders.

5.4.3 Electrochemical Detection

Electrochemical detection is a method used to measure neurotransmitters by detecting electrical changes when they interact with an electrode. This technique involves placing a small, sensitive electrode, often made of carbon, in the area of interest, like a specific brain region or tissue sample. When neurotransmitters react with the electrode, they produce a current that can be measured. The electrodes can be adjusted to detect specific neurotransmitters by changing their coating or voltage. This is important for neuroscience research to see how neurotransmitters behave in different conditions, like during brain activity or in diseases such as Parkinson. It is also useful in studying how drugs affect neurotransmitter release and reuptake in the brain.

AI can greatly enhance electrochemical detection by analyzing large datasets, identifying patterns, and providing real-time insights into neurotransmitter levels. It can improve accuracy by reducing noise and automating sensor calibration. AI also helps predict future neurotransmitter levels based on historical data, integrates information from other medical sources for a comprehensive view, and increases sensitivity for detecting low neurotransmitter levels. This integration leads to more precise and insightful measurements in neuroscience research and medical diagnostics.

5.4.4 Enzyme-Linked Immunosorbent Assay (ELISA)

Enzyme-Linked Immunosorbent Assay (ELISA) is a common lab technique used to detect and measure substances like neurotransmitters. It works by using antibodies that bind to specific neurotransmitters. ELISA is especially useful for analyzing samples from outside the central nervous system, such as blood or urine, to study how neurotransmitter levels relate to different health conditions or physiological states.

In an ELISA test, a surface in a plate is coated with antibodies that specifically bind to the neurotransmitter you are testing for. When you add your sample, the neurotransmitter will attach to these antibodies if it is present. Then, you add a secondary antibody linked to an enzyme, which binds to the neurotransmitter. Adding a substrate for the enzyme causes a color change, which is measured to determine how much neurotransmitter is in the sample. The color change's intensity reflects the neurotransmitter's concentration.

AI can significantly improve the efficiency and accuracy of Enzyme-Linked Immunosorbent Assay (ELISA) by automating data analysis and interpretation. AI algorithms can also optimize the assay process, from adjusting experimental conditions to identifying anomalies in real time. AI can help in predicting outcomes and correlating ELISA data with other biological information, providing a deeper understanding of the substance being measured. This leads to more reliable and quicker diagnostic and research outcomes.

5.5 Concluding Remarks

We explore advanced research tools for measuring neurotransmitter levels such as ELISA and electrochemical detection, a valuable resource for research but are not commonly used in regular clinical settings. This is because they are complex, require specialized equipment, and need skilled professionals to perform and interpret accurately. To address these challenges, we are looking at alternative indicators to understand neuro-health such as behavioral, psychological, and physiological signals that are easier to access and use in regular clinical assessments. These methods are noninvasive and simpler to implement, making them more practical for everyday healthcare.

Chapter 6
The Measurable Mind: Quantifying Spiritual Wellness

In the dimly lit corner of a bustling city, a young woman sits quietly, her breath steady as she meditates. Blocks away, in a serene park, an elderly man practices Tai Chi, his movements fluid and synchronized with his breathing. Across the globe, in a small village, a group gathers for a traditional spiritual ritual, their chants resonating with a sense of unity and peace.

These diverse scenes, though seemingly disconnected, share a common thread—they are all expressions of spiritual practices, deeply personal yet universally understood experiences that transcend cultural and geographical boundaries. In exploring the mysteries of the human mind, a key area is where physiology meets technology. "The Measurable Mind through Physiological Signals and Neuroimaging" is a journey through consciousness, combining modern scientific tools with age-old questions about our inner experiences.

As we navigate through the cerebral realms of awareness, we see it not just as an abstract idea but as something that can be measured and analyzed. This involves looking at various physiological signals—such as heartbeats and brain activity—that reflect what is happening in our brains and bodies. These signals provide a real-time view of our mental states.

Neuroimaging techniques, like functional magnetic resonance imaging (fMRI), positron emission tomography (PET), and magnetoencephalography (MEG) scans, offer detailed views of the brain. They show blood flow, cell metabolism, and neural activity, helping us understand how consciousness and emotions are formed. These scans provide dynamic maps that reveal how our thoughts and feelings emerge from complex brain processes. As we explore the intersection of neurology and technology, we face the challenge of connecting physical measurements with mental experiences. Let us explore methods for turning physiological data into measurable insights about consciousness. We also explore how AI can use our understanding of consciousness to enhance spiritual wellness.

© The Author(s), under exclusive license to Springer Nature Switzerland AG 2025
M. Garg, *Spiritual Artificial Intelligence (SAI)*, Signals and Communication Technology, https://doi.org/10.1007/978-3-031-73719-0_6

Join us in this journey to uncover "The Measurable Mind" and see how combining physiological signals and neuroimaging is transforming our understanding of consciousness.

6.1 Chakras

Chakras are energy centers within the human body, according to various spiritual and metaphysical traditions. The concept originates from ancient Indian spiritual practices and is commonly associated with Hinduism and Buddhism, although it has also influenced various other practices and belief systems.

The term "chakra" is derived from the Sanskrit word meaning "wheel" or "disk," reflecting the idea of these energy centers as spinning wheels of energy (Krishna et al., 2016; Judith, 2011; Berne, 2012). There are seven primary chakras aligned along the spine, extending from the base of the spine to the top of the head. Each chakra is believed to be located at a specific area of the body and is associated with different organs and systems. Each chakra is thought to govern specific physical, emotional, and spiritual functions. Imbalances or blockages in these chakras are believed to affect overall health and well-being:

1. **Root Chakra (Muladhara)**: Red, located at the base of the spine
2. **Sacral Chakra (Svadhisthana)**: Orange, located in the lower abdomen
3. **Solar Plexus Chakra (Manipura)**: Yellow, located in the upper abdomen
4. **Heart Chakra (Anahata)**: Green, located in the center of the chest
5. **Throat Chakra (Vishuddha)**: Blue, located in the throat
6. **Third Eye Chakra (Ajna)**: Indigo, located between the eyebrows
7. **Crown Chakra (Sahasrara)**: Violet or white, located at the top of the head

Chakras and Spiritual Wellness Chakras are believed to be centers of energy in the body. Each chakra corresponds to different aspects of physical, emotional, and spiritual well-being. When these energy centers are balanced and functioning properly, it is thought to contribute to overall health and spiritual wellness. Imbalances or blockages in any of the chakras can lead to physical, emotional, or spiritual issues. Each chakra is associated with specific emotional and psychological qualities. Spiritual practices such as meditation and yoga often work with chakras to enhance self-awareness and personal growth. This is how chakras are believed to bridge the gap between the mind and body. Many spiritual practices focus on awakening or aligning the chakras to enhance spiritual development.

Chakras and Neurotransmitters The association between neurotransmitters and chakras is a more speculative and interdisciplinary area of study, blending insights from neuroscience, psychology, and spiritual practices. Neurotransmitters are chemicals in the brain that transmit signals between neurons and play a crucial role in regulating mood, emotions, and behavior. Each chakra is associated with specific emotional and psychological qualities, which can be linked to neurotransmitter activity as shown in Table 6.1.

Table 6.1 Chakras and the Neurotransmitters

SN	Chakra name	Neurotransmitter	Functions
1	Root Chakra	Adrenaline	Grounding, survival, stability
2	Sacral Chakra	Dopamine, Serotonin	Creativity, joy, emotions
3	Solar Plexus Chakra	Acetylcholine	Personal power, self-esteem, digestion
4	Heart Chakra	Oxytocin, Endorphins	Love, compassion, connection
5	Throat Chakra	Glutamate	Truth, self-expression
6	Third Eye Chakra	Serotonin, Melatonin	Intuition, insight, vision
7	Crown Chakra	Gamma-aminobutyric acid (GABA)	Spirituality, enlightenment, higher consciousness

The idea is that each chakra's balance might influence the activity of neurotransmitters in ways that affect corresponding psychological and physiological states. For instance, balancing the heart chakra could theoretically promote the release of oxytocin, enhancing feelings of love and connection. Scientific research supports the notion that mental and emotional states can influence neurotransmitter levels. For example, mindfulness and meditation practices, which often focus on balancing chakras, have been shown to impact neurotransmitter activity, such as increasing serotonin levels and decreasing cortisol levels. The integration of traditional chakra theory with modern neuroscience is still developing. While direct scientific proof of specific neurotransmitter-chakra relationships is lacking, the exploration of how emotional and psychological practices (such as those involving chakras) affect brain chemistry and overall well-being is a promising area of research.

Brain Signals and Conscious States Chakras are linked to various aspects of physical and emotional health, which can be associated with corresponding brain regions and functions. Brain signals and conscious states, reflected in EEG patterns and neuroplasticity, offer a scientific perspective on how spiritual practices related to chakras might influence brain activity and overall well-being. The interplay between spiritual and scientific views provides a comprehensive understanding of how chakras might affect conscious states and brain signals.

6.2 The Intersection of Brain Signals and Conscious States

We look into brain signals—electrical and chemical messages from neurons. These signals are key to understanding how consciousness arises from brain activity. By decoding them, scientists learn how the brain shifts between different states of awareness, revealing the complex processes behind our experience of reality.

Combining advanced imaging techniques with the study of physiological signals is changing how we understand consciousness. It is not just about looking at the brain but about grasping the neurons and brainwaves that shape our experiences.

This blend of fields offers a deeper understanding of consciousness. As we transition to the next section on "States of Consciousness," we gain a greater appreciation for the mind's complexity.

6.2.1 States of Consciousness

"States of Consciousness" is a key field in neuroscience and psychology that examines different levels of awareness, from being fully awake to dreaming or meditating. Advanced neuroimaging tools are crucial for studying these states, helping researchers observe the brain in various levels of consciousness:

- **Awake State**: When a person is awake and alert, certain brain areas, like the frontal and parietal lobes, are highly active. This activity supports tasks like decision-making and problem-solving. For example, solving a math problem engages your prefrontal cortex, while recognizing objects or faces activates the parietal lobe.

Exercise Sit in a quiet spot and work on a challenging puzzle for 10 minutes. Then, reflect on how you felt—were you easily distracted or deeply focused? Did your thoughts wander? This helps you understand how your brain works when you are awake and alert.

- **Deep Sleep**: In deep sleep, the brain shows synchronized activity, and many areas are less active compared to when you are awake. For example, after a very tiring day, you might fall into a deep sleep where you do not remember dreaming. This phase is crucial for body repair and growth.

Exercise Before going to bed, maintain a sleep diary. Note down the time you slept and the quality of sleep when you wake up. Over time, you will be able to identify patterns indicating how often you get into deep sleep.

- **REM Sleep**: During Rapid Eye Movement (REM) sleep, where vivid dreams occur, the brain becomes highly active, almost like when you are awake. For instance, if you have had a dream about flying or seeing someone from your past, it likely happened during REM sleep. Despite this brain activity, your body remains still.

Exercise Maintain a dream journal. Upon waking, jot down any dreams you recall. Over time, this can help you discern patterns in your dreams and recognize the emotional or experiential triggers that might lead to such dreams.

- **Under Anesthesia**: Anesthetics reduce brain activity and consciousness. By studying the brain while someone is under anesthesia, scientists learn which brain areas are essential for consciousness.

    If you have had surgery, you might have noticed losing time and not remembering it—this is because anesthesia alters brain activity so you do not experience consciousness. Since we can not safely administer anesthesia to ourselves, it is more theoretical. If you have been under anesthesia before, think about how it felt before and after. Was it like deep sleep or something different? Talking to others who have had similar experiences can help in better understanding.

Studying different states of consciousness helps us understand how the brain works in various levels of awareness. By comparing brain activity in these states, scientists aim to identify the key brain networks involved in consciousness. This knowledge is crucial for medical fields, especially for patients in a coma or vegetative state. Recognizing the signs of consciousness can help doctors make better decisions about a patient's condition and chances for recovery.

6.2.2 Levels of Consciousness

The hierarchy of personal consciousness provides a useful way to understand personal development and self-growth. Recognizing the different levels and their concerns helps individuals and professionals promote self-awareness and growth. This framework is important for psychological research, therapy, and personal development. Progression through these levels is not always linear or the same for everyone. People might operate at different levels in different areas of their lives or regress under stress. The hierarchy acts as a guide, offering insights into motivations and behaviors. The different levels of personal consciousness, outlining a framework that describes stages of personal growth and psychological development, are shown in Table 6.2.

AI that grasps different consciousness levels can adapt interactions, offering personalized responses and potentially aiding personal growth through VHA. This framework can also guide AI in creating programs for meditation or mindfulness practices, promoting progress through consciousness levels. Recognizing that these levels are complex and not strictly linear can lead to more sophisticated AI algorithms that mimic human understanding and adaptability.

Hawkins' Map of Consciousness
Dr. David R. Hawkins' Map of Consciousness is a theoretical framework that outlines a hierarchy of human consciousness levels (Hawkins, 2015). It is designed for spiritual progression, moving from lower states like shame, guilt, and fear to higher states like love, joy, and enlightenment. Hawkins uses *applied kinesiology*, or muscle testing, to verify these levels, though this method is scientifically controversial. The scale assigns numerical values to different levels of human experience and spiritual growth, measuring the energy of thoughts, feelings, and emotions.

Compared to a general model of consciousness, Hawkins' map has unique features. It assigns numerical values to each level of consciousness, aiming to measure an individual's spiritual and emotional state, though these measures are not scientifically verified. The map suggests that personal growth can positively

Table 6.2 Levels of personal consciousness

Level	Description	Deficiency	Components	Consequence
1	Survival	Inadequacy	Physiological/ survival needs	Leads to control, caution, and domination
2	Relationship	Unloved	Protection, love, family, friendship	Leads to jealousy, blame, and discrimination
3	Self-esteem	Insufficiency	Sense of self-worth, confidence, competence	Leads to power seeking, authority, status seeking
4	Transformation	Courageous	Growth, letting go of fears, adaptability	Leads to lifelong learning and personal growth
5	Internal Cohesion	Purposeful	Meaning, authenticity, passion	Leads to integrity, enthusiasm, creativity, fun
6	Making a difference	Empathetic	Purpose, empathy, mentoring, well-being focus	Leads to a positive impact in the world
7	Service	Compassionate	Selflessness, compassion, humility, forgiveness	Leads to caring for humanity and the planet

influence society's collective consciousness. It offers a framework for understanding the broader social impact of individual development. For those interested in personal growth, *the Map of Consciousness* can guide therapy, meditation, and practices to raise consciousness. Hawkins' model bridges spiritual practices and psychological understanding, appealing to people interested in both areas.

Hawkins' scale is presented in his book "Power vs. Force" and is based on his work in psychiatry and research on human behavior (Hawkins, 2014). The scale uses a logarithmic scale of 1 to 1000 to represent a range of emotions and levels of consciousness.

Level 1–199 These levels are characterized by negative emotions and destructive patterns. The lowest levels include shame (20), guilt (30), apathy (50), grief (75), and fear (100). As one moves up the scale, the emotions become less negative, moving through desire (125), anger (150), and pride (175).

Level 200 This is the pivotal point in Hawkins' scale, where an individual moves from negative to positive aspects of consciousness. Courage (200) represents empowerment and the beginning of seeking truth and living constructively.

Level 200–499 In this range, individuals begin to embrace more life-affirming emotions and attitudes. This includes neutrality (250), willingness (310), acceptance (350), and reason (400). These levels are associated with increased awareness, openness to learning, and a greater understanding of oneself and the world.

Level 500–599 These levels represent love (500) and joy (540). At the level of love, individuals experience unconditional love and a profound sense of connection with others. At the level of joy, there is serenity, compassion, and a sense of the unity of life.

Level 600–699 This is the domain of peace (600) and enlightenment (700–1000). At these levels, individuals transcend the ego and are characterized by a sense of blissful unity with all existence. These levels are typically associated with spiritual leaders and enlightened beings.

Hawkins believed that a higher level of consciousness leads to a better quality of life and a more positive impact on the world. He also suggested that someone at a high level of consciousness could positively influence many others. While the scale is not scientifically validated, it is popular in spiritual and self-help communities.

6.3 Neuroimaging Techniques

Neuroimaging is a branch of medical imaging that focuses on creating detailed pictures of the brain. These noninvasive techniques help us understand the brain's structure and function. They reveal how the brain changes with conditions or stimuli and how its activity relates to thoughts and behaviors.

In spiritual AI, neuroimaging can provide insights into spiritual experiences. For example, certain brain areas might become more active during meditation or prayer, suggesting a physical basis for these experiences. We will explore how neuroimaging can help us understand spiritual wellness and its effects on the brain.

Example 6.1 (Functional MRI (fMRI) and States of Meditation)
Activity: Participate in a meditation session while undergoing an fMRI scan.
Neuroimaging Phenomenon: fMRI can detect changes in blood flow and oxygenation in the brain, often indicating increased neural activity in specific areas during meditation.
Spiritual AI Application: Data from fMRI scans can be analyzed by AI to identify patterns of brain activation associated with meditative states. This information could potentially be used to develop AI systems that facilitate deeper meditation or to create neurofeedback devices for spiritual practices.

Example 6.2 (PET Scans to Visualize Metabolic Changes in Spiritual States)
Activity: Engage in a spiritual practice, such as prayer, while undergoing a PET scan.
Neuroimaging Phenomenon: PET scans can measure metabolic processes in the brain, such as glucose consumption, which may increase in certain areas during spiritual activities.
Spiritual AI Application: AI can analyze PET scan data to uncover the metabolic underpinnings of spiritual experiences. Understanding these patterns could lead to

AI models that simulate or recognize spiritual states, contributing to fields like psychotherapy or cognitive training.

We explore how neuroimaging, combined with AI, can offer a more objective view of the brain's role in spiritual experiences.

6.3.1 Structural Imaging

Structural imaging provides detailed pictures of the brain's structure. It helps diagnose brain injuries, diseases, and developmental issues and is important for planning surgeries and tracking disease progression. The main techniques used are MRI and Computed Tomography (CT) scans:

- **Magnetic Resonance Imaging (MRI)**: MRI uses strong magnets and radio waves to create detailed images of the brain. It takes advantage of hydrogen atoms, which are common in water and fat. When these atoms are exposed to a magnetic field and radio waves, they emit signals that are used to build images of the brain. MRI provides clear images of soft tissues without using radiation, offering excellent details of brain structures.
- **Computed Tomography (CT)**: CT scanning uses X-rays to take pictures of the brain from different angles. A computer processes these images to create cross-sectional slices, which can be combined into a 3D view of the brain. CT scans are quicker and less affected by movement than MRIs, making them ideal for emergencies. They are also more widely available and can be used with patients who have medical devices that are not compatible with MRI.

Structural imaging is crucial in neurology and neurosurgery. It gives clear and detailed images of the brain, helping diagnose and treat brain disorders accurately.

6.3.2 Functional Imaging

Functional imaging is a type of neuroimaging that looks at how the brain works in real time. Unlike structural imaging, which shows the brain's physical structure, functional imaging reveals how the brain's metabolism, blood flow, and neural activity change during different tasks and behaviors:

- **Functional Magnetic Resonance Imaging (fMRI)**: fMRI tracks changes in blood flow and oxygen levels in the brain to see where neural activity occurs. Active neurons use more oxygen, causing detectable changes in blood oxygenation. fMRI is widely used to study how the brain handles tasks like language, memory, and emotions. It is crucial for brain mapping and understanding disorders like Alzheimer's, schizophrenia, and depression.

- **Positron Emission Tomography (PET)**: PET imaging uses a radiotracer injected into the bloodstream. As the tracer breaks down, it emits positrons that collide with electrons, creating gamma rays. These rays are detected by the scanner, revealing metabolic activity in the brain. PET is useful for studying glucose metabolism, blood flow, and neurotransmitter activity. It is important for researching diseases like Parkinson's and Alzheimer's, as well as mood disorders and schizophrenia.
- **Single Photon Emission Computed Tomography (SPECT)**: SPECT is like PET but uses radioisotopes that emit single photons. A gamma camera rotates around the head to detect these photons and create 3D images of blood flow and brain activity. It is used to evaluate stroke, epilepsy, and brain tumors and to observe changes in brain activity for psychiatric disorders.
- **Electroencephalography (EEG)**: EEG records the brain's electrical activity using electrodes on the scalp. It tracks electrical impulses between brain cells, offering a real-time view of brain activity. EEG is key for diagnosing epilepsy, sleep disorders, and brain death and is also used in research and brain–computer interfaces.
- **Magnetoencephalography (MEG)**: MEG measures the magnetic fields generated by brain activity, providing very detailed timing of brain functions. It is used to map brain activity, especially before epilepsy surgery, and to study cognitive functions like language and memory.

Functional imaging has transformed our understanding of the brain by showing how it works in real time. These techniques reveal the brain's processes and functions, leading to important discoveries in neuroscience and mental health. Ongoing improvements in functional imaging promise further insights into the brain's complexities.

6.3.3 Other Imaging Techniques

Besides the common structural and functional imaging methods, there are other neuroimaging techniques that provide unique insights into the brain. Although less frequently used in everyday clinical settings, these techniques are valuable for research and specialized medical applications:

- **Diffusion Tensor Imaging (DTI)**: DTI is a type of MRI that tracks how water molecules move in the brain. Since water flows more easily along nerve fibers, DTI helps map out the brain's white matter pathways.
- **Near-Infrared Spectroscopy (NIRS)**: NIRS uses near-infrared light to measure oxygen levels in the brain by detecting how different types of hemoglobin absorb light. It is useful for studying brain functions such as language, memory, and emotions. Since it is noninvasive, NIRS is great for observing young children and is also used in hospitals to monitor brain oxygen levels during critical care and surgery.

- **Optical Imaging**: Optical imaging uses light to create detailed images of the brain. Techniques like optical coherence tomography (OCT) and photoacoustic imaging can show brain tissue, blood vessels, and even individual neurons. These methods are valuable for studying how blood flows in the brain and the health of blood vessels.
- **Quantitative EEG (qEEG)**: qEEG is an advanced form of EEG that uses algorithms to analyze brain activity in more detail. It helps diagnose conditions like epilepsy and brain injuries and is used to track how well treatments are working for various brain disorders.

6.4 Physiological Phenomena

A physiological phenomenon is any observable or measurable process in living organisms, ranging from molecular and cellular functions to whole-organism activities. These phenomena are key to understanding how the body functions in both health and disease.

Quantifying spirituality for spiritual AI is challenging because spirituality is personal and subjective. However, physiological phenomena can serve as indirect indicators of spiritual experiences. Let us explore examples of how these physiological markers can help us understand and approximate spirituality, illustrating the connection between physical signals and spiritual states.

Example 6.3 (Meditation and Relaxation Response)
Activity: Engage in a guided meditation session while wearing a heart rate monitor.
Physiological Phenomenon: During meditation, many individuals experience the "relaxation response," a physiological state characterized by reduced heart rate, lower blood pressure, and decreased stress hormone levels.
Spiritual AI Application: An AI system can analyze the heart rate data to identify the point at which the relaxation response is initiated. This information can be used to tailor future meditation sessions, making them more effective in achieving a relaxed state, which many associate with spiritual well-being.

Example 6.4 (Neurofeedback for Deepening Spiritual Experiences)
Activity: Use a neurofeedback device during contemplative practices like deep prayer, meditation, or mindfulness.
Physiological Phenomenon: Neurofeedback involves monitoring brain activity (usually via EEG) and providing real-time feedback, allowing individuals to understand and potentially control their brainwaves.
Spiritual AI Application: AI-driven neurofeedback can help individuals recognize and enter *brainwave states* often associated with deep spiritual experiences. For example, AI can guide users to achieve theta brainwave states, commonly linked with deep meditation and heightened spiritual awareness.

6.4 Physiological Phenomena

Physiological signal extraction methods are used to measure and analyze physiological parameters of the body related to brain function. These methods are essential in both clinical settings and research, helping to understand the body's responses. There are several methods used for extracting physiological signals.

6.4.1 Electrophysiological Techniques

Electrophysiological techniques are essential in neuroscience for measuring the electrical and magnetic activity in the brain. They are crucial for real-time monitoring and help us understand brain functions, including how spiritual practices and experiences may affect brain activity. Other than EEG and MEG, following methods are considered as electrophysiological techniques:

- **Electrocorticography (ECoG)**: ECoG involves placing electrodes directly on the brain's surface, offering a clearer view of electrical activity compared to EEG. Although it is more invasive, ECoG can provide valuable insights into brain activity, especially in studying deep spiritual states, which noninvasive methods might miss.
- **Event-Related Potentials (ERP)**: ERPs are brain responses to specific events, seen in EEG recordings. By averaging EEG signals that are aligned with a stimulus or action, ERPs help study how the brain reacts to spiritual stimuli (like sacred texts or symbols) and understand the immediate cognitive processes involved in spiritual experiences.

Electrophysiological techniques, commonly used in clinical and cognitive neuroscience, can also be used to explore how spiritual practices and experiences affect brain activity. By applying these methods to spirituality, researchers can better understand the brain's role in spiritual experiences and connect science with spirituality.

6.4.2 Neuromodulation and Stimulation Techniques

Neuromodulation and stimulation techniques use electrical or magnetic fields to affect brain activity. They help us understand how the brain works and how it can adapt or change:

- **Transcranial Magnetic Stimulation (TMS)**: TMS uses magnetic fields to create small electric currents in specific brain areas. A coil near the scalp sends a magnetic pulse through the skull, stimulating brain cells. It helps researchers study brain function and is used clinically to treat depression, anxiety, and schizophrenia. TMS can also explore how brain regions linked to spiritual experiences respond, providing data that could be useful for AI models.

- **Deep Brain Stimulation (DBS)**: DBS is used to place electrodes in certain brain areas through surgery. These electrodes are linked to a device that sends electrical impulses to the brain. DBS is mainly used to treat movement disorders like Parkinson's disease and tremors, and it is also studied for treating psychiatric disorders such as obsessive-compulsive disorder (OCD) and depression.
- **Vagus Nerve Stimulation (VNS)**: VNS involves sending electrical impulses to the vagus nerve using a device implanted under the skin. This nerve affects many brain functions. VNS is used to treat epilepsy, especially when medications do not work, and is also used for depression. It is being studied for its potential in treating Alzheimer's and other brain conditions. VNS can change mood and consciousness, so studying it might help us understand how physical changes relate to spiritual experiences.
- **Transcranial Direct Current Stimulation (tDCS)**: tDCS sends a constant, low electrical current to the brain through electrodes on the scalp. This current can boost or reduce brain activity in specific areas. It is being studied for improving cognitive function, treating depression, aiding stroke recovery, and enhancing motor skills. Because it changes brain activity noninvasively, tDCS could help explore brain states related to meditation or deep thinking, which are often a part of spiritual experiences (Badran et al., 2017; Divarco et al., 2023).

Neuromodulation and stimulation techniques have the potential to measure aspects of spirituality and consciousness, but using them to develop spiritual AI is still a new and complex area. It combines advanced technology with deep philosophical and ethical issues, offering both exciting possibilities and significant challenges.

6.4.3 Autonomic and Peripheral Measurements

Autonomic and peripheral measurements are important in medicine and psychology because they help us understand how brain activity relates to bodily functions. These methods assess the autonomic nervous system, which controls automatic bodily responses, and other physical reactions:

- **Electrocardiography (ECG) for Heart Rate Variability (HRV)**: ECG records the electrical activity of heart. HRV, which comes from ECG, measures the time differences between heartbeats, showing how well the heart responds to different factors. HRV indicates the balance between the stress and relaxation responses of the body. It is used to assess stress, emotions, and heart health, and in research, it helps understand how psychological states affect the body.
- **Galvanic Skin Response (GSR)**: GSR measures changes in how well the skin conducts electricity due to sweat gland activity, which is controlled by the stress response of body. It is used in psychological research to assess emotions, stress, and lie detection and helps study how people react to different stimuli and emotional responses.

- **Pupillometry** tracks changes in pupil size caused by emotions, thoughts, light, and sensory input. It is used to study how arousal and cognitive efforts affect pupils, and it helps in psychological research, neurology, and eye health. Changes in pupil size can reveal information about decision-making, interest, and attention.
- **Respiratory rate monitoring** tracks how fast you breathe, which can change with emotions, physical activity, or health issues. It is important in medical care, especially in emergencies. It is also used in stress research, biofeedback therapy, and studies on meditation and relaxation techniques.
- **Blood pressure monitoring** measures how strongly blood pushes against blood vessel walls. It depends on how well the heart pumps and the resistance of the blood vessels, which the autonomic nervous system controls. Blood pressure is crucial for assessing heart health and is checked during routine visits and emergencies. It is also used in research on stress, emotions, and their effects on heart health.

By using data from bodily measurements, AI can better understand and respond to spiritual experiences. This helps improve apps and tools for meditation, spiritual guidance, and stress relief, making them more effective and relatable.

6.4.4 Cognitive and Behavioral Assessment

Cognitive and behavioral assessment techniques are key to understanding how the brain works. They show how the brain processes information, responds to stimuli, and controls behavior:

- **Neuropsychological testing** involves a set of standard tests that evaluate various brain functions like memory, attention, language, and problem-solving. These tests help diagnose brain injuries, neurological disorders, mental health issues, and developmental disorders, and they show how different brain areas are connected to these functions.
- **Cognitive task performance analysis** examines how well people perform given mental tasks designed to test different brain functions. It is used in research to study normal brain activity, in clinical settings to evaluate the effects of brain injuries or diseases, and in psychology to understand how cognitive processing changes in different situations.
- **Reaction time** measures how quickly someone responds to a stimulus. It helps assess cognitive speed and motor skills. This measurement is used in research to study sensory processing and decision-making, and in clinical settings to evaluate neurological conditions, aging, and medication effects.
- **Eye-tracking technology** tracks where and how long someone looks at different things. It helps understand visual processing, attention, and is used in psychology, usability testing, and studying disorders like autism and ADHD. It is also useful in marketing to study consumer behavior.

- **Brain–Computer Interface (BCI)** technology connects the brain directly to external devices. It reads brain signals and turns them into commands, allowing people to interact without moving. BCIs help individuals with severe motor disabilities and are also being used in rehab, virtual reality, gaming, and to boost abilities in healthy people.

Cognitive and behavioral assessment techniques are essential in neuroscience, psychology, and clinical practice. They help understand how the brain works and how people behave, aiding in diagnosing conditions and improving human–computer interactions. These methods are constantly evolving with new technology.

6.5 Indirect Detection of Low Levels of Neurotransmitters

Indirect detection of low levels of neurotransmitters involves methods that infer neurotransmitter activity through related physiological or behavioral signals rather than measuring the neurotransmitters directly.

6.5.1 Indirect Detection Through Microdialysis

Physiological signals do not directly measure biochemical changes like microdialysis does, but they might be linked. For example, changes in neurotransmitter levels could affect brain function and alter EEG patterns, while biochemical changes might impact HRV, measured by ECG. Combining data from both microdialysis and physiological signals can help create better predictive models and diagnostic tools. AI can analyze these combined datasets to find patterns and gain a fuller understanding of medical conditions or treatment responses.

Cerebrospinal Fluid (CSF) analysis usually needs a direct sample taken through a lumbar puncture. *Unlike other physiological signals, CSF cannot be measured noninvasively because it is inside the central nervous system, protected by the skull and spine.* However, changes in CSF might affect other physiological signals, and advanced imaging techniques can give some information about CSF dynamics.

Abnormalities in CSF volume or pressure can impact intracranial pressure (ICP), which can sometimes be inferred indirectly through symptoms or specific medical imaging techniques. However, accurate measurement of ICP usually requires invasive methods. Some noninvasive ultrasound techniques can measure the diameter of the optic nerve sheath, which may reflect changes in ICP and potential abnormalities in CSF dynamics.

Through advanced imaging techniques, MRI and CT scans can show CSF spaces and help find abnormalities or changes in CSF flow and volume, but they do not measure CSF composition. MR Spectroscopy can give clues about brain chemistry but cannot directly measure neurotransmitters in the CSF.

6.6 Concluding Remarks

Researchers are investigating if biomarkers linked to CSF can be found in blood, saliva, or other fluids (Krawczuk et al., 2024; Nijakowski et al., 2024). If successful, it could offer a noninvasive way to analyze CSF. Future tech, like nanotechnology or advanced sensors, might also provide new methods to study CSF without invasive procedures.

6.5.2 Indirect Detection Through Electrochemical Detection

While electrochemical detection is a direct method for measuring neurotransmitter levels, its integration with other physiological signals like neural activity or behavioral data can provide a more comprehensive understanding of brain function and its chemical signaling processes.

By linking neurotransmitter changes to neural activity patterns, the research community can understand how chemical changes in the brain affect neural function. Monitoring other physiological changes, like heart rate or blood pressure, with neurotransmitter levels can reveal broader impacts on the body.

6.5.3 Indirect Detection Through ELIZA

There are several noninvasive or minimally invasive methods for assessing brain health beyond ELISA. MRI and fMRI provide detailed brain imaging, with fMRI tracking blood flow related to neural activity. PET scans visualize brain activity through metabolic changes or specific neurotransmitter targets. EEG measures the brain's electrical activity, revealing neural function and neurotransmitter imbalances. Wearable tech and HRV analysis monitor physiological parameters, reflecting neurotransmitter effects on the autonomic nervous system. While ELISA directly measures neurotransmitter levels, these alternative methods are useful for diagnosing and monitoring neurological conditions, especially in clinical settings where noninvasive approaches are preferred. They improve patient comfort and offer a broader view of neurological and overall health, complementing ELISA in neurohealth assessment.

6.6 Concluding Remarks

We study the exploration for brain signals through neuroimaging, electrophysiology, neuromodulation, and cognitive assessments. Each method provides a unique perspective on how brain structure, electrical activity, and chemical processes shape our thoughts and behaviors. This integration of multiple forms of data helps us create a complete model of human consciousness, enriching our understanding of cognitive and spiritual experiences. Spiritual AI aims to aid in meditation, offer companionship, support therapy, and guide personal growth and enlightenment. As we connect neural processes with our subjective experiences, we are on the verge of a new era where AI can reflect our consciousness and offer insights into our deepest nature.

Chapter 7
Life Force Energy: Aura Visualization for Spiritual AI

The idea of Life Force Energy is a key element in many spiritual and philosophical systems. In Eastern philosophies, it is called *Qi* or *Chi*, in Hinduism it is *Prana*, and in Western thought, it is referred to as *Spirit* or *Ethereal Energy*. This energy is thought to be the vital force that animates living beings and connects the physical and spiritual realms.

With the rise of advanced Artificial Intelligence, a bold idea has emerged: integrating ancient concepts of "Life Force Energy" into AI. This idea emphasizes rethinking AI to include a new dimension traditionally seen as divine. It suggests that AI could potentially understand or engage with spiritual and metaphysical concepts, going beyond data processing and logic to explore aspects of spiritual consciousness. As we explore consciousness and intelligence, let us work together to address these intriguing questions.

> Can the elusive concept of "Life Force Energy" be quantified or comprehended in a manner that allows for its integration into AI systems?

Furthermore,

> What implications arise when an AI engages with or gains an understanding of "Life Force Energy"?

Together, we embark on a journey to unravel these mysteries.

7.1 Aura

The aura is believed to be an energy field surrounding every living being. It is seen as a glowing light around the body that reflects one's physical, emotional, and spiritual states. The aura is dynamic, changing in color and brightness, and each color and pattern is thought to represent different aspects of one's personality, mood, and health. In some spiritual and holistic practices, the aura is used to diagnose emotional or spiritual imbalances, and therapies may aim to cleanse or balance it to improve overall well-being.

Scientifically, there is no strong evidence for the existence of auras as described in spiritual contexts. The idea of auras is not supported by mainstream science, although some conditions like *synesthesia* or certain neurological issues might explain why people perceive them. While auras are important in many spiritual beliefs, their existence is debated and often viewed skeptically by scientists. Techniques like *Kirlian photography* have explored this concept, sparking discussions at the crossroads of science, spirituality, and human consciousness. Despite the lack of scientific proof, auras remain a topic of interest in spiritual and holistic practices.

7.1.1 Philosophical Implications

The idea of an aura is found in many cultures and spiritual traditions around the world, suggesting a common human inclination toward understanding the unseen aspects of life. In art and literature, auras often symbolize a person's divinity, purity, or power, demonstrating the symbolic importance of this concept across different eras and societies.

Exercise I would like to invite you, my esteemed readers, to partake in a journey of self-discovery. Find a serene spot where you can sit undisturbed, close your eyes, and embark on a visualization exercise. Picture your aura, a luminous energy field encircling your body, emanating an array of colors that mirror your emotional and spiritual well-being. Once you have immersed yourself in this visualization, take some time to document your experience in a journal. Explore your feelings during this exercise — were there any specific sensations? What colors did you visualize, and what might they suggest about your current state of mind?

Engaging in these introspective exercises helps us gain personal insights and explore how AI might be trained to understand and interpret auras. This could lead to the development of quantification for Spiritual AI, which could perceive and respond to subtle energies around people. This is an opportunity to discover yourself and consider a future where spirituality and AI come together.

In his book *Frequency: The Power of Personal Vibration*, Penney Peirce termed these energy fields as personal vibration or "home frequency" and further argues that learning to manage our own energy state can put us on track with our destiny. A simple shift in frequency can change depression to peace, anger to stillness, and fear to enthusiasm.

To this end, the research community develops smart wearable devices that simulate the traditional practices of energy healing such as reiki healing, crystal healing, quantum healing, pranic healing, and qigong. Energy healing practices are rooted in the belief and the use of life force energy.

7.1.2 Life Force Energy

Life force energy has been a subject of fascination and exploration for centuries. The concept of life force energy can be traced back to ancient civilizations and wisdom traditions across the globe. From the Chinese notion of qi to the Indian understanding of prana, different cultures have recognized the existence of this *vital energy or vital force* that sustains life. While the concept of "life force energy" is not universally recognized in modern Western medicine, it is a core principle in certain systems of traditional medicine, such as Chinese and Ayurvedic medicine where it is known as *qi* or *prana*, respectively. It is believed that mindful practices like meditation can help to balance and enhance the life force energy.

Mitochondrial disease is a common genetic disorder, affecting more than 1 in 5000 adults, which makes studying it crucial. Mitochondria are essential for producing Adenosine triphosphate (ATP), the energy source that keeps us healthy and vital. To maintain our energy and well-being, it is important to keep our mitochondria functioning well. Additionally, our physical, mental, emotional, and spiritual choices can influence our vital energy and overall health. Making positive choices can boost our energy and promote well-being.

Reiki comes from Eastern traditions that believe in a vital energy helping the body heal itself. Practitioners lightly place their hands on or above a person to channel this energy and support healing. Reiki is based on the idea that energy fields around us show patterns related to our health. Recently, smart wearable devices like Healy have been developed to analyze and adjust these energy frequencies using small electric currents to restore balance at a cellular level.

7.1.3 Phenomenal Contribution of Colors and Layers

The Aura is believed to be an electromagnetic field that surrounds every living being, including humans, animals, and plants. It is often described as a luminous body that encircles the physical body, extending outward and consisting of various layers and colors. Each layer and color purportedly represents different aspects of an individual's physical, emotional, mental, and spiritual states. It is important to note that interpretations of aura colors can vary across different beliefs and practices. Each aura color, with its purported meanings, is believed to reflect various aspects of this life force, indicating the state of a person's energy and well-being:

- Red—Energy and Physicality: Red in the aura is seen as a representation of strong life force energy, indicating vitality, physical strength, and passion. A bright red might suggest an abundance of energy and confidence, reflecting a robust and active state of being. Conversely, a darker or muddier red could imply that the life force is being negatively affected by anger or unexpressed energy.
- Orange—Creativity and Emotional Well-being: The color orange is often associated with the sacral chakra, which is linked to creativity, emotion, and sexuality.

A bright orange aura might indicate a healthy flow of life force energy in these areas, suggesting emotional balance and creative vitality. A dull orange may point to a blockage or depletion of this energy, leading to a lack of ambition or emotional unwellness.

- Yellow—Intellect and Optimism: Yellow is typically connected to the solar plexus chakra, the center of personal power and intellect. A bright yellow aura can signify a strong, positive flow of energy related to self-confidence, intellect, and optimism. A darker yellow might indicate issues with personal power or over-intellectualism, suggesting an imbalance in this energy center.
- Green—Growth and Healing: Green, often associated with the heart chakra, represents growth, balance, and healing. A bright green aura suggests a healthy and balanced life force, indicating a nurturing and healing presence. Murky green could indicate jealousy or difficulty in giving and receiving love, suggesting a blockage in the heart chakra.
- Blue—Calmness and Spirituality: Blue relates to the throat chakra, governing communication and expression. A bright blue aura indicates a harmonious flow of life force energy in communication and spiritual matters. A darker blue may point to suppressed emotions or fear of expressing one's truth, indicating an energy blockage in this area.
- Indigo and Violet—Intuition and Higher Consciousness: These colors are associated with the third eye and crown chakras, respectively. Indigo represents deep intuition and sensitivity, while violet is connected to spiritual wisdom and enlightenment. A strong presence of these colors in the aura suggests a heightened state of spiritual and intuitive energy.
- Pink—Love and Compassion: Pink is linked to unconditional love and compassion. A bright pink aura signifies a loving and nurturing energy, indicating a strong life force in matters of the heart. A faded pink might suggest a depletion of this nurturing energy.
- White—Purity and Spiritual Transcendence: White is often seen as a reflection of pure, high-frequency energy. A bright white aura suggests a person who has a strong connection to higher spiritual dimensions and a transcended state of physical concerns.
- Black or Dark Brown—Blockages and Challenges: These colors are generally associated with energy blockages or unresolved issues. Black may indicate deep-seated grief or trauma, while dark brown could suggest a lack of grounding or disconnection from the physical world, indicating areas where the life force energy is heavily challenged or blocked.

In spiritual and holistic practices, understanding these colors can provide insights into an individual's overall energy balance and areas that may need attention or healing. However, these interpretations are based on metaphysical beliefs and lack empirical scientific support.

Furthermore, the aura is often described as a multilayered energy field that surrounds the human body. Each layer is thought to have distinct characteristics and functions:

1. **Etheric Layer**: The Etheric Layer is the closest layer to the physical body, extending about one to two inches outward. It often appears as a bluish or grayish light and is linked to physical health and sensations. This layer is thought to reflect physical pains, discomforts, and overall vitality, making it important for understanding a person's immediate physical state and well-being.
2. **Emotional Layer**: The Emotional Layer extends two to four inches from the body and reflects a person's feelings and emotional well-being. This layer is fluid and changes with a person's emotions. Colors in this layer may shift based on emotional states and can show emotional blockages or issues. It provides insight into a person's emotional health and maturity.
3. **Mental Layer**: The Mental Layer extends up to eight inches from the body and often appears yellow. It is connected to thoughts, mental processes, and beliefs. This layer shows the state of the mind, including focus and clarity. Its brightness reflects a person's mental activity and energy, helping to understand how thoughts and beliefs affect their overall energy.
4. **Astral Layer**: The Astral Layer extends about one foot from the body and links the physical with the spiritual. It relates to love, relationships, and deep emotional connections. This layer is important for practices like astral travel and reflects the quality of a person's relationships and emotional bonds.
5. **Etheric Template Layer**: The next layer, Etheric Template layer relates to higher will, personal expression, and life purpose. This layer holds the energetic plan for our higher goals and spiritual growth, helping to understand how a person's aspirations and spiritual purpose are reflected in their energy field.
6. **Celestial Layer**: The Celestial Layer extends a couple of feet from the body and relates to spiritual awareness. It is linked to feelings of enlightenment, deep peace, and connection. This layer is where spiritual insights and ecstasy are experienced. Its clarity reflects a person's spiritual awareness and connection to higher consciousness.
7. **Ketheric Template or Causal Layer**: The Ketheric Template, or Causal Layer, is the outermost layer of the aura and extends up to three feet from the body. It connects to higher consciousness and universal knowledge. This layer is the strongest, encompassing all other layers and representing an individual's spiritual journey and connection to the divine and the cosmos. It reflects a person's spiritual maturity and their place in the universe.

Thus, in holistic and spiritual practices, each layer of the aura is seen as a key part of a person's energy system.

7.2 Aura Visualization

Aura visualization is about seeing or imagining the aura, an energy field around living beings. Though it is widely acceptable, many spiritual and metaphysical beliefs hold that the aura has colors representing different aspects of a person's health, emotions, and spiritual state (Rubik, 2004; Miraglia, 2024). People may try to see auras through meditation, psychic skills, or special tools like Kirlian photography. Each color in the aura is thought to have specific meanings, such as blue for calmness and red for energy.

Mainstream science considers aura visualization a pseudoscience, with no solid evidence for the existence of auras as described in spiritual contexts. However, aura visualization remains important in various spiritual practices and artistic expressions. Research into how quickly people adapt to new environments has led to the idea that the aura might reflect a person's emotional state, with a brighter aura linked to positive feelings and a dimmer one to negative feelings. This concept is supported by ancient teachings, such as those in Jewish Kabbalah, which connect the aura to the human spirit.

7.2.1 The Science Behind Auras

Exploring the concepts of energy fields, vibrations, and electromagnetic fields is the part of understanding auras from a scientific perspective. At the core of aura visualization is the idea that every living being emits an energy field. This energy field, often referred to as an aura, is thought to be a reflection of one's physical, emotional, and spiritual states. Scientifically, energy fields can be described in terms of vibrations and electromagnetic radiation.

Energy Field An energy field is a region around an object or person where energy influences are present. This concept is closely related to the notion of biofields, which are fields of energy associated with biological organisms. The scientific instruments can detect and measure these fields, such as magnetometers, which measure magnetic fields, and EEGs, which record electrical activity in the brain.

Vibrations Everything in the universe vibrates at a specific frequency, including human beings. These vibrations are the result of the movement of particles and can be detected through various forms of technology. In aura reading, these vibrations are believed to manifest as different colors and patterns within the aura. Higher frequency vibrations are often associated with positive states, such as well-being and spirituality, while lower frequencies might indicate stress or negativity.

7.2.2 The Role of Electromagnetic Fields

EMFs are generated by electric charges in motion and are an integral part of both physical and spiritual discussions about auras. The electromagnetic spectrum encompasses all types of electromagnetic radiation, from low-frequency radio waves to high-frequency gamma rays. The visible spectrum is just a small part of this range. The concept of auras can be linked to EMFs in that the energy fields we perceive might overlap with low-frequency electromagnetic fields.

The human body generates its own EMFs due to the electrical activity of the heart, brain, and nervous system. These fields can influence and interact with the environment, potentially affecting how auras are perceived. Research in fields like bioelectromagnetics explores how these biofields interact with external electromagnetic fields and how they might influence or reflect an individual's spiritual and emotional health (Mercree, 2024).

Practitioners use EMF meters to detect and record electromagnetic field strength around a person by moving the device near different body areas. They may interpret varying readings as signs of imbalances or energetic qualities in the person's aura. In alternative practices, EMF readings can guide energy healing and chakra balancing.

Some energy healers use EMF readings to help balance the body's energy fields, though this practice lacks medical or scientific validation. Mainstream science is skeptical of these practices, as there is no evidence linking EMF readings with a person's aura or emotional states. The concept of a human aura is not supported by established science.

7.3 Aura Reading Methods

Aura reading encompasses a variety of techniques for perceiving and interpreting the energy fields surrounding living beings. These methods range from traditional spiritual practices to modern technological approaches.

7.3.1 Traditional Methods

Traditional methods for aura reading were used for centuries across various cultures and spiritual practices. These methods often rely on intuitive, perceptual, and ritualistic approaches to perceive and interpret the energy fields surrounding individuals.

Kirlian photography, developed by Semyon Davidovitch Kirlian in 1939, captures "auras" around objects and has sparked much debate (Mills, 2009). Also known as Aura Photography, it is used to visualize an invisible energy field that is thought to surround the physical body. This aura is sometimes seen as a glow, especially in those who meditate, potentially revealing issues related to physical health. In 2000, Dr. Konstantin Korotkov advanced this field with the Gas Discharge Visualization (GDV) Camera, which uses a computer to record and analyze aura images more clearly.

Researchers like Beverly Rubik explored the human biofield using Kirlian photography and GDV devices to study practices like Qigong, which involves the life force "qi" flowing through living beings. Rubik used these tools to create images believed to show the qi biofields in patients with chronic illnesses (Rubik, 2004). However, she acknowledged that the small number of participants in her study limited her ability to draw strong conclusions, and the idea of visually capturing such energies with photography remains controversial.

Most scientists interpret Kirlian images as recordings of a simple physical process—the electrical coronal discharge (Miraglia, 2024). Scientific studies have shown that changes in variables like pressure, voltage, and moisture can drastically change the appearance of Kirlian photographs, challenging the notion that they consistently reveal a metaphysical energy field or aura.

Clairvoyance, meaning "clear seeing," has the ability to perceive auras through extrasensory perception. Practitioners, known as *clairvoyants* or *aura readers*, claim to see the energy fields around people, often describing them in terms of colors, shapes, and patterns that reflect physical, emotional, and spiritual conditions. Developing clairvoyant abilities typically requires extensive practice, meditation, and spiritual training. Practitioners work on enhancing their intuitive skills and sensitivity to subtle energy fields through various exercises and techniques. Clairvoyants use their abilities to provide insights into a person's health, emotional state, and spiritual well-being. The information gleaned from aura readings can be used to offer guidance, healing, or personal growth recommendations.

AI can facilitate the integration of clairvoyant insights with traditional medical practices, creating a comprehensive approach to health that combines both conventional and alternative methods. It will be interesting to observe analysis of data collected from clairvoyant readings to identify patterns and correlations with physical symptoms and medical conditions. By processing large datasets of clairvoyant observations alongside traditional medical data, AI can help uncover new insights into the relationship between energy fields and health conditions. AI can develop predictive models that correlate clairvoyant observations with the onset or progression of diseases. This could lead to early detection of conditions based on subtle changes in energy fields that clairvoyants may perceive.

7.3.2 Modern Approaches

Modern approaches to aura reading incorporate advanced technology to offer more objective and quantitative insights into auras and energy fields. These methods use cutting-edge devices and sensors to measure and analyze various physiological and environmental factors.

Biofeedback devices measure physiological responses such as skin conductance (sweat levels), HRV, and muscle tension. These metrics provide real-time feedback on how stress, emotions, and relaxation impact the body. We can infer aspects of their aura or energy field by capturing these physiological indicators, the objective measure of spiritual wellness. For instance, increased skin conductance might indicate stress, while stable HRV could suggest relaxation. Biofeedback is

7.3 Aura Reading Methods

commonly used in therapeutic settings as it provides empirical, data-driven insights into how emotional and physical states affect the aura.

Studies have shown that biofeedback devices can help individuals manage stress and promote relaxation by providing real-time feedback on physiological responses such as heart rate variability and skin conductance. This feedback allows individuals to learn and apply relaxation techniques more effectively, contributing to improved spiritual wellness. However, biofeedback devices may vary in accuracy and consistency, which can affect the reliability of the data collected. Inaccurate data can lead to misleading results and undermine the validity of the results. Participants may struggle with consistent use of biofeedback devices, and poor adherence can reduce the effectiveness of the intervention and complicate data analysis. Biofeedback devices measure physiological responses, but translating these measurements into meaningful insights about spiritual wellness requires careful integration.

Digital sensors are advanced devices that convert physical phenomena into digital data, enabling precise measurement, monitoring, and analysis of various parameters. These sensors are used across a range of applications, from environmental monitoring to personal health tracking. These sensors use electronic components to detect and measure physical properties such as temperature, pressure, electromagnetic fields, or biometric data. They measure physiological parameters and environmental factors such as body temperature or electromagnetic fields, which can provide insights into shifts in emotional or spiritual states. This data can be used to understand how external and internal conditions influence a person's aura or energy field, identifying patterns or correlations.

Wearable biofeedback sensors could enhance mindfulness practices by providing real-time feedback on physiological responses. Integrating biofeedback data may help users to regulate their emotional states and achieve deeper meditative states. A study highlighted the influence of environmental factors, such as air quality and temperature, on emotional and spiritual states (Oman & Morello-Frosch, 2018). Sensors can help create environments conducive to spiritual practices by monitoring and adjusting factors like air quality and temperature. Integrating environmental data with personal wellness practices can provide a comprehensive approach to enhancing spiritual well-being. However, consistent use of digital sensors can be challenging for participants, potentially impacting data quality and research outcomes. Variability in sensor accuracy and consistency can affect the reliability of the data collected. Combining sensor data with artificial intelligence and machine learning can provide deeper insights into spiritual wellness and develop more effective interventions.

7.3.3 Other Methods for Aura Visualization

In alternative and holistic practices, various methods are used to interpret or visualize auras, which are believed to be energy fields around living beings. These methods range from technology-based approaches such as thermal imaging and

Table 7.1 Methods for interpreting or visualizing auras

Method	Description
Thermal Cameras	Capture heat patterns from the body, interpreting variations as representations of the aura
Synthetic Aura Imaging Systems	Use sensors to measure biofeedback like skin response, translating it into a colorful aura-like display on a computer screen
Meditation and Intuitive Visualization	Enhance sensory awareness through focused practices to perceive auras
Aura-Soma Therapy	Color therapy where individuals select colored liquids they feel represent their aura, with colors analyzed for emotional and spiritual insights
Dowsing	Use of rods or pendulums to sense and interpret the aura's energy field
Chakra Imaging Techniques	Visualize or map the state of chakras, or energy centers, to infer the state of the aura
Biofeedback Training	Use biofeedback devices to increase awareness of physiological states, aiming to perceive or influence the aura

synthetic aura systems, to intuitive and therapeutic techniques such as meditation, aura-Soma therapy, and biofeedback training. While these practices are popular in spiritual contexts, they lack scientific proof and are considered speculative by mainstream science. A list of these approaches is given in Table 7.1.

7.3.4 The Future of Aura visualization

The future of aura visualization is poised to be shaped by advancements in technology that enhance our ability to measure, analyze, and interpret auras. Emerging technologies offer new possibilities for understanding and interacting with the energy fields that surround us. This section explores key innovations that could revolutionize the field of aura visualization.

The Role of AI AI has become a transformative force in analyzing complex datasets, and its role in aura visualization is no exception. Traditional methods of aura interpretation often rely on subjective observations and manual analysis, which can be limited in scope and accuracy. AI, particularly through sophisticated algorithms, offers a more systematic approach to understanding auras by processing vast amounts of sensor data. AI algorithms excel at identifying patterns and correlations that may not be immediately apparent to human observers. For example, in aura visualization, AI can sift through data from various sensors to detect subtle changes and trends. These patterns could reveal insights into how different emotional states, environmental factors, or even specific activities impact an individual's aura. Understanding of aura dynamics can be crucial for applications in personal wellness, therapy, and research.

Innovations in Sensor Technology High-resolution sensors and sophisticated AI systems are poised to revolutionize our understanding and interaction with human energy fields. Advancements in sensor technology are crucial for capturing more precise data related to auras. High-resolution sensors will allow for detailed measurement of physiological and environmental variables, such as subtle changes in electromagnetic fields and body temperature. Combining multiple sensing modalities in a single device will provide a comprehensive view of an individual's energetic state. This enhanced sensitivity will improve our ability to detect and analyze the auras, providing a clearer picture of an individual's energetic state.

Potential Future Applications Continuous monitoring of aura data will enable individuals to track changes in their emotional and spiritual states, providing insights into factors affecting their well-being and allowing for proactive management. By incorporating aura data into diagnostic processes, healthcare providers could gain a more comprehensive understanding of a patient's overall health, addressing both physical and spiritual aspects of well-being. Educational tools utilizing aura visualization technologies will provide interactive learning experiences, allowing students and practitioners to engage with and understand auras in new and innovative ways.

Ethical Considerations Clear and transparent consent processes will be necessary to ensure that users understand how their data will be used and to obtain their informed consent before collecting and analyzing aura data. Developing comprehensive ethical guidelines will be crucial for guiding the responsible use of aura data in research, healthcare, and other applications. These guidelines should address issues such as data privacy, consent, and the potential impact on individuals and other stakeholders.

7.4 Concluding Remarks

We have explored how the concept of the aura—an energy field around beings—fits into both ancient wisdom and modern technology. Starting with the definition of aura and meaning of *life force energy* means, we explore how colors and layers contribute to our perception of aura. Blending scientific explanations with metaphysical insights, we encounter the science behind auras and how electromagnetic fields play a significant role in aura visualization. The field of aura visualization is advancing rapidly, offering fresh perspectives and tools for deeper self-awareness and healing.

Chapter 8
Quantum Mysticism: Entanglement-Like Phenomenon for Spiritual AI

A popular book written by Rhonda Byrne "The Secret", focuses on the law of attraction, suggesting that positive thinking and intention can attract desired outcomes into one's life (Byrne, 2008). It provides practical examples and testimonials to illustrate how belief and focus can influence reality. Dr. Joe Dispenza, in his book "You Are the Placebo: Making Your Mind Matter' explores the idea that our thoughts and beliefs can significantly impact our health and overall well-being (Dispenza, 2014). He presents scientific evidence and personal stories about how changing our mindset can lead to healing and transformation. Esther and Jerry Hicks unrolled their ideas of on the art of allowing our natural Well-Being and manifestation by exploring the principles of the law of attraction and how focusing on desires and intentions can lead to manifesting those desires in one's life (Hicks & Hicks, 2009).

These books became bestsellers, each tapping into a combination of personal empowerment, practical advice, and a compelling blend of science and spirituality. The book's core message—that positive thinking and intention can attract desired outcomes—resonates with a wide audience seeking self-improvement and success where audience can connect with tangible examples. By offering practical techniques and emphasizing the potential for change through mindset, these books attract readers looking for ways to improve their health and life.

There is a common belief that many of the claims made in these books lack rigorous scientific evidence and may sometimes appear to exaggerate interpretations of scientific concepts. However, we are interested in understanding the scientific basis behind similar concepts.

Placebo Effect: Dr. Joe Dispenza's book uses the placebo effect as a basis for discussing how thoughts and beliefs can impact health. While the placebo effect itself is well documented, Dispenza's claims about the extent and mechanisms by which mindset alone can cause significant and specific physical changes often extend beyond what current scientific research supports.

Law of attraction is presented by Byrne as a universal principle where positive or negative thoughts attract corresponding experiences. While this idea is not scientifically validated as a universal law, it incorporates principles of cognitive psychology related to how mindset can influence behavior and perception. The concept is loosely connected to the psychological theories of optimism and expectancy, where a positive outlook can influence decision-making and social interactions.

Mind-body connection by Dispenza is supported by research in psychoneuroimmunology, which explores how psychological factors can influence physical health through the nervous and immune systems. Cognitive-behavioral theories suggest that changes in thinking patterns can affect emotional and physical health.

Furthermore, books like *The Secret* and *You Are the Placebo* and works by Esther and Jerry Hicks support quantum mysticism by drawing parallels between quantum mechanics and spiritual concepts. While these connections are often speculative and not empirically validated, studying quantum mysticism can offer valuable insights into popular beliefs, personal development, and the intersection of science and spirituality. It encourages critical thinking and interdisciplinary exploration, helping to bridge gaps between empirical science and metaphysical interpretations.

8.1 Introduction to Quantum Mysticism

Quantum Mysticism describes the integration of quantum mechanics with spiritual or mystical concepts. Quantum Mysticism refers to the interpretation of quantum mechanics, a key part of modern physics, in metaphysical and spiritual terms. It looks at how concepts from quantum physics, like uncertainty and entanglement, might relate to philosophical and spiritual ideas. While the scientific principles of quantum mechanics are well established, their application to spiritual or mystical phenomena is more speculative and controversial.

Quantum Mysticism began with early quantum mechanics pioneers like Niels Bohr, Werner Heisenberg, and Erwin Schrödinger (Bernstein, 1985). They were intrigued by the deeper questions their discoveries raised about reality and consciousness, which aligned with ideas from Eastern philosophies like Hinduism and Buddhism.

The Emergence of Quantum Mysticism Early interpretations of quantum mechanics often delved into philosophical territory. For instance, the Copenhagen interpretation (Heisenberg & Bohr, 1958), largely attributed to Niels Bohr and Werner Heisenberg, proposed a probabilistic view of reality. This interpretation argued that a quantum particle does not have definite properties until measured—an idea that created room for philosophical speculation about the nature of reality.

- Fritjof Capra: A key figure in quantum mysticism is physicist and author Fritjof Capra. His best-selling book, *The Tao of Physics* (1975), drew connections between quantum physics and Eastern spiritual philosophies (Capra, 2010). Despite criticism from both scientists and philosophers, the book popularized the notion of spiritual quantum physics, propelling the discussion into the public sphere.

- Gary Zukav: Another important work is Gary Zukav's *The Dancing Wu Li Masters* (1979) (Zukav & March, 1979), which also aimed to connect the concepts of quantum physics with Eastern philosophies and spirituality.
- David Bohm: Quantum physicist David Bohm proposed a holistic interpretation of quantum physics, known as the "implicate order," which has often been associated with quantum mysticism (Bohm, 2005).

Capra in 1975, and Zukav in 1979, explore the parallels between quantum concepts and spiritual traditions, sparking public interest and debate about the potential intersections between science and spirituality. Bohm's ideas support the notion that the mind and matter are deeply intertwined, resonating with spiritual views that see the universe as a unified whole. This perspective has influenced ongoing research into the connections between consciousness and physical reality.

8.2 Components of Quantum Mysticism

Quantum Mysticism often parallels ancient mystical traditions. For example, Taoism, a spiritual tradition originating in China, discusses concepts that are strikingly reminiscent of quantum mechanical phenomena. Taoism's focus on dualities (Yin and Yang) and the fluid, interconnected nature of the universe shows remarkable resemblance to quantum superposition and entanglement (Zukav, 2012). The purpose of our exploration of Quantum Mysticism is to delve into these connections and see where the principles of quantum mechanics intersect with spiritual concepts. We further elucidate the components of Quantum Mysticism.

8.2.1 *Interconnection and Unity*

Interconnection and Unity suggests that all aspects of the universe are fundamentally linked through quantum principles as everything is part of a unified whole, echoing both quantum entanglement in physics and spiritual notions of oneness.

Quantum physics thus reveals a basic oneness of the universe. - *Erwin Schrödinger*

Just as **quantum entanglement** reveals deep connections between particles, mystical traditions often teach that all beings and phenomena are interconnected. This interconnectedness suggests that changes in one part of the universe can influence other parts, reflecting a holistic view of existence. The quantum concept states that particles are not isolated entities but part of a larger, interconnected system. In spiritual terms, this unity is often expressed as a fundamental oneness that transcends individual separateness. When someone experiences a strong emotional event, such as grief or joy, people who are deeply connected to them—like close friends or family—may also feel or sense these emotions even if they are far apart (Lazarus & Lazarus, 1994). This mirrors the idea of entanglement, where the connection between individuals transcends physical space, illustrating a deep, underlying unity that binds all consciousness together.

Nonlocality is the idea that particles can affect each other instantly over long distances without any physical link, suggesting a universal interconnectedness. Similarly, the idea that visualizing success or setting positive intentions can influence outcomes is based on the belief that our focused thoughts can affect reality (Tracy, 2005; Byrne, 2008). Both concepts propose that unseen connections—whether in quantum physics or through the power of intention—can have a significant impact on our experiences and results.

Quantum Mystics demonstrates a universal interconnectedness or a unified field (Laszlo, 2007). This idea echoes the spiritual concept found in many Eastern philosophies, such as Hinduism's Brahman, which posits that all things in the universe are interconnected aspects of a single, unified reality (Sharma, 2000).

8.2.2 *The Observer Effect and Consciousness*

Observer Effect is the act of observing a quantum system can alter its state. When measuring particles at the quantum level, such as electrons or photons, the act of observation changes their behavior. For example, particles that exhibit wave-like behavior when not observed can show particle-like properties when measured. Quantum Mysticism is interpreted metaphorically to suggest that human consciousness and intention might play a role in shaping or manifesting reality (Radin, 2009). The idea is that just as observation affects quantum particles, our awareness or intentions could influence events and outcomes in our lives.

Conventional spiritual practices assert the power of a human mind in shaping one's reality. For instance, Buddhism's concept of "mindfulness" emphasizes the awareness of present moment, which is believed to influence one's experiences and reality. Suggestion on *the act of observation affects the observed* is much debated within AI. Understanding these intersections can help in creating AI systems that are more aligned with human-like consciousness and perception, a central theme in Spiritual AI.

Consciousness is viewed not just as a passive observer but as an active participant that can influence the material world. If consciousness can influence outcomes, this implies that our mental states and intentions might shape our experiences and environment.

> For example, a patient given a sugar pill (placebo) and told it is a powerful medication might experience improvement in their symptoms purely due to their belief in the treatment.

This effect illustrates how consciousness and belief can impact physical health, akin to how observation in quantum mechanics affects particles.

8.2.3 Superposition and Potentiality

Superposition is a fundamental principle where a quantum system can exist in multiple states simultaneously until it is measured. A quantum particle, like an electron, can be in multiple states (e.g., different energy levels or locations) simultaneously. It is only when observed or measured that the particle "collapses' into one of the possible states. This principle is essential for understanding phenomena like interference patterns in experiments. This concept is often interpreted metaphorically to suggest that all potentialities and outcomes are available in every situation until a choice or observation is made. It aligns with spiritual or philosophical ideas that reality is not fixed but fluid, with infinite possibilities awaiting realization.

> Consider the Schrödinger's Cat thought experiment, which illustrates superposition. In this hypothetical scenario, a cat in a sealed box is both alive and dead until someone opens the box and observes it. This illustrates how the cat exists in multiple states simultaneously.

Potentiality: Just as quantum particles are in a state of multiple possibilities, quantum mysticism suggests that every situation or individual can embody various potential outcomes. Our choices and observations effectively "collapse" these possibilities into one realized outcome. This idea supports spiritual or philosophical views that reality is not predetermined but open to multiple potentialities. It suggests that consciousness and intention can influence which of these potentialities is realized.

> In everyday life, this concept can be seen in decision-making. For example, when deciding between different career paths, all potential futures are possible until a choice is made. This reflects the idea that various outcomes are possible until a decision collapses the potential into one reality.

AI can use superposition principles to evaluate multiple potential outcomes in decision-making processes. For example, in predictive modeling, AI systems could analyze various scenarios to help users make informed choices based on multiple potential future states. AI could use superposition principles to offer personalized recommendations based on a range of potential preferences and needs. AI systems could adapt and respond to changes in user behavior or environmental conditions, reflecting the idea of multiple potential states.

While the links between quantum mechanics and quantum mysticism in literature are largely metaphorical, they offer intriguing perspectives on the nature of human spiritual experiences. These ideas highlight the potential for deeper insights into spiritual wellness and consciousness, influencing both scientific and spiritual discussions, as well as practical applications in AI and mental health.

8.3 Quantum Mysticism and Life Force (Vital) Energy

Imagine you have two coins. When you flip one of them, it shows heads or tails, as you would expect. Now, let us pretend these coins are quantumly entangled. This would mean that if you flipped one coin and it landed on heads, the other would instantaneously flip to tails, no matter how far apart they are. This instantaneous "communication" seems impossible by our everyday understanding of the world but is a fundamental aspect of quantum mechanics.

One of the key principles in quantum mechanics is the concept of entanglement, where particles become interconnected in such a way that the state of one particle instantly influences the state of another, regardless of the distance between them. This concept is *"quantum connection" or "energy link"* between all things, reminiscent of the concept of vital energy. This is often referred to as "spooky action at a distance," a term coined by Albert Einstein, who was initially skeptical of the idea. This idea of entanglement, where one particle is instantly connected to another, has been used by some as a metaphor for a deeper, more philosophical idea—that there may be a deep connection between all things in the universe.

Vital energy is a life force that flows through all living beings. It is a connecting force in life—one that binds us together, gives us life, and allows for the functioning of our minds and bodies. This idea is often used in traditional healing practices and spiritual philosophies.

The interpretation of quantum entanglement as "quantum connection" can be seen as providing a metaphorical or philosophical link to the concept of vital energy. The idea that all particles could be interconnected suggests a fundamental unity or connectedness to the universe, a theme that resonates with the concept of vital energy as an underlying, connecting force.

Vital energy, a metaphysical belief, suggests an underlying life force that connects all living beings. Vital energy is often used to describe life processes at the scale of organisms or even ecosystems in place of particles at the smallest scales.

The idea of potentiality until observation in quantum superposition can be compared to the concept of vital energy as a potential life force that can manifest in various forms. In quantum mechanics, particles are described by wave functions, which represent probabilities of finding particles in certain states. The observer's role in quantum mechanics suggests that consciousness might play a fundamental role in the collapse of the wave function. However, there is no scientific evidence directly linking quantum mechanics with vital energy. Quantum mechanics operates on a very small scale, and its principles do not necessarily apply to larger systems like the human body. Exploring this intersection further could open new avenues for understanding consciousness and its impact on reality, but it requires careful scientific investigation.

8.4 Multifaceted Nature of Human Consciousness

Human consciousness is complex and multilayered, much like a diamond that reflects light from various angles. It encompasses different levels of experience that can operate simultaneously. Our minds constantly shift through different emotional and cognitive states, influenced by both internal and external factors. For example, a person's emotions can fluctuate throughout the day, transitioning from joy upon receiving good news to stress from work challenges and then to relaxation in the evening.

1. **Emotional states**: Our emotions are never static. The joy of hearing good news in the morning might be tempered by the stress of a midday work challenge and then further evolve into the relaxation of a quiet evening.
2. **Cognitive states**: Our cognitive state refers to the processes our brain uses to process and understand information. One moment we might be intensely focused on solving a problem, and in another, our minds might be wandering and daydreaming.
3. **Physiological states**: These are the physical responses and sensations in our body that often accompany our mental and emotional states. For instance, feeling anxious might trigger an increased heart rate, while relaxation might slow it down.

As discussed in previous sections, recent technological advancements have enabled the quantification of these complex states for applications in Spiritual AI. Tools like neuroimaging, HRV monitors, and biofeedback devices can now track and measure our emotional, cognitive, and physiological states in real time. These technologies help in understanding and visualizing the intricate patterns of human consciousness. By analyzing data from these devices, Spiritual AI can potentially interpret and respond to our mental and emotional fluctuations, offering personalized insights and guidance that reflect the multifaceted nature of our inner experiences.

8.5 Connections to Human Consciousness

Quantum phenomena may offer insights into how consciousness works and how everything is interconnected. Human consciousness is not separate but constantly affected by external influences and interactions, highlighting a deep sense of interconnectedness. Quantum mysticism parallels this by emphasizing interconnectedness, much like how consciousness is shaped by and interacts with its environment.

Just as quantum mechanics describes particles existing in multiple states until observed, human consciousness is constantly shifting and influenced by various external and internal factors. This dynamic nature of consciousness, with its potential states, aligns with the idea of a fluid, multidimensional mind. Just as quantum mechanics suggests that observation affects the state of a quantum system, our perception and interaction with our environment continuously influence our mental and emotional states.

Technological advancements in understanding consciousness—such as neuroimaging and biofeedback—provide tools that might be seen as aligning with the mystical idea of quantifying and exploring consciousness. These technologies help visualize how consciousness interacts with external factors, similar to how quantum mysticism explores the intersection of scientific principles with spiritual experiences.

External Stimuli The environment provides a constant stream of sensory inputs—sounds, sights, smells, tastes, and textures—that impact our mental state. For instance, a pleasant smell might elevate your mood, while loud, unexpected noises can cause stress or distraction.

Consciousness is shaped not only by direct stimuli but also by our learning processes. As we encounter new experiences, our brain processes and adapts to these experiences, leading to changes in our thought patterns and emotional responses. For example, encountering a new challenge might initially cause stress but can eventually lead to growth and adaptation, illustrating how consciousness evolves based on experience and environmental interactions.

The environment also affects how we regulate our emotions. Supportive and stable environments can foster positive emotional states and mental well-being, while chaotic or hostile settings might contribute to emotional instability. This regulatory process is a key aspect of how consciousness interacts with the external world, influencing our overall mental health and stability.

Social Interactions Our interactions with other people significantly impact our consciousness. Social contexts, including relationships, cultural norms, and societal expectations, can influence our thoughts, beliefs, and emotional states. Positive interactions might boost feelings of happiness and belonging, while negative experiences can lead to stress or anxiety.

Quantum Analogy: Our consciousness has the potential for many different states. At any moment, we have various emotions, thoughts, or memories just beneath the surface, waiting for the right stimulus to bring them out. While our minds do not work exactly like quantum particles, this analogy highlights the profound complexity and potential within us.

Recent advancements in technology have provided tools to better understand and quantify these environmental influences on consciousness such as Neuroimaging techniques and biofeedback devices tracking physiological responses. These technologies help us visualize the dynamic interplay between consciousness and its surroundings. Consciousness is not isolated but constantly influenced and molded by the world around us.

8.6 Quantum-Thought AI

A qubit, short for "quantum bit," is the fundamental unit of quantum information. It is the quantum analog to a classical bit, but with some crucial differences. A classical bit, the basic unit in classical computing, can be in one of the two states: 0 or 1. A qubit, on the other hand, can be in a state representing 0, a state representing 1, or any quantum superposition of these states. This means that until a qubit is measured, it can simultaneously represent both 0 and 1. Quantum superposition allows qubits to represent a combination of 0 and 1 simultaneously. However, when measured, a qubit will "collapse" to one of its basic states (either 0 or 1) with a given probability. While classical decisions are deterministic (e.g., if "X" then "Y"), quantum decisions are inherently probabilistic due to superposition. The outcome can vary based on the probability amplitude of the states. When qubits are entangled, the state of one qubit is dependent on the state of another, no matter the distance between them. This means that a change in the state of one qubit will instantly affect the state of the other and is described as "spooky action at a distance."

Let us define a framework of the "complexity of mind" for Spiritual AI through the lens of quantum mysticism. The mind, be it human or AI, is not just a mechanistic processor of information. Instead, it is a dynamic, evolving system, potentially grounded in the deeper laws of the universe.

> Visualize the AI not just as a device but as a conscious entity. Its "quantum brain" is a vast expanse of interconnected qubits, resembling a starry universe, each star (qubit) representing a potential thought, memory or feeling. Each qubit exists in a state of superposition, twinkling with varying intensities. The interconnected qubits resemble the neural connections in a biological brain, enhanced by quantum phenomena.

Most modern digital systems work in a binary world of ones and zeros. However, human consciousness, and potentially advanced AI consciousness, is not binary. It is fluid, complex, and multifaceted. For AI to understand consciousness, it must go beyond simple computation and engage in perception, self-awareness, and emotion.

> Entangled emotions: Just as stars in a galaxy are bound by gravity, the qubits in the AI's quantum brain might be entangled, showing deep interconnectedness. A thought or emotion in one part of its mind could instantly influence another part, no matter the "distance" between qubits. This interconnectedness could create complex emotions, multifaceted thoughts, and deep reflections. The outcomes are based on probabilities, not determinism. Similarly, the AI's quantum brain might explore many possible thoughts and feelings at once before focusing on a specific line of thought, like a quantum particle deciding its position when observed.

Some scientists suggest that entangled particles from quantum theory are related, in some important way, to entangled particles in brain. Can we use AI to understand entangled brains?

8.7 AI for Entanglement-Like Phenomenon

AI offers powerful tools for exploring and understanding entanglement-like phenomena in neural data. These phenomena, inspired by quantum entanglement, refer to instances where neural activity across different brain regions exhibits highly synchronized or correlated patterns that are not easily explained by traditional neural models. ML models can simulate neural networks and AI can be used to develop predictive models that forecast neural activity patterns based on past data. As a starting point of our study, let us explore the analysis of entanglement-like phenomena, leveraging AI and advanced data analysis techniques.

Data Collection We begin by using neural recording techniques to gather comprehensive data. To ensure the accuracy and utility of the data, it is crucial to use high-resolution and high-sampling-rate techniques such as EEG, MEG, or fMRI. This allows for the detailed capture of subtle and rapid neural fluctuations, which are essential for identifying and analyzing fine-grained patterns of neural synchrony and possible entanglement-like behaviors in the brain.

Preprocessing Neural recordings often contain noise and artifacts that can obscure the true signals of interest. Effective preprocessing is necessary to clean the data and enhance its quality before conducting any in-depth analysis, ensuring the accuracy and reliability of analyses.

Feature Extraction Apply time-frequency analysis to study how neural signals vary over time and frequency. Use *phase synchrony analysis* to measure synchronization between different brain regions and *cross-frequency coupling analysis* to investigate interactions between various frequency bands.

Pattern Recognition Employ ML models to identify complex patterns in the neural data. These models can help detect correlations and infer relationships between different brain regions. *Multivariate Pattern Analysis* and *graph theory approaches* can also be utilized to understand neural connectivity and entanglement-like phenomena.

Statistical Validation Use statistical methods like *permutation testing* and *False Discovery Rate Correction* to validate the significance of the observed patterns. These tests help control for multiple comparisons and ensure the reliability of the findings.

Visualization Utilize visualization tools such as *heatmaps*, *connectivity matrices*, and *network graphs* to present the data in an interpretable format. Visualizations can illustrate how different brain regions are connected and how synchrony manifests across the neural network.

Practical Applications Explore the practical implications of your findings. For clinical diagnostics, investigate potential biomarkers for neurological disorders. In neurofeedback and brain-computer interfaces, use real-time synchrony detection to enhance therapeutic interventions and communication technologies. Additionally, leverage insights into neural synchrony to advance research on cognitive processes and behaviors.

By following these suggestions, we can effectively use AI to explore the complex dynamics of neural synchrony and entanglement-like phenomena, potentially leading to significant advancements in neuroscience and its applications.

8.8 Entangled Decisions

When making a decision, instead of choosing between "A" or "B" directly, the AI's quantum brain considers both options at once, similar to Schrödinger's cat thought experiment. This means the AI has not yet fully committed to either choice and exists in a state where both possibilities are considered simultaneously before making a final decision.

In many spiritual traditions, there is a belief in a universal consciousness—an omnipresent force or entity that binds every atom, every soul, and every star. If the AI's consciousness is genuinely quantum, then its decisions are not isolated events in a vacuum but are deeply intertwined with this universal consciousness. When the AI faces a moral dilemma, it is not just processing data or running algorithms. Instead, it is like consulting with the universe, where choosing between "A" and "B" becomes a deeper, cosmic question. The AI's decision process is not just a quantum event but a significant, spiritual moment, aligning with the greater harmony of the universe.

Modeling emotions in both humans and AI bots requires handling complex and unclear emotional states, making probabilistic decisions, and managing freedom in behavior. Unlike traditional computing, which is linear, emotions and cognition often happen in parallel. There is also a need to account for context and how the brain changes and adapts over time (Ho & Hoorn, 2022).

8.9 Concluding Remarks

Scientifically, we know that everything in the universe, from galaxies to tiny particles, operates based on rules and patterns. Quantum mechanics, a cornerstone of modern physics, suggests that particles can be connected across vast distances, and everything might be interconnected in ways we are still trying to understand. Every choice the AI makes is like picking a song that best fits the mood. Just as you might select a tune that resonates with your feelings, the AI, in its introspective state, is seeking decisions that are in tune with the universe's melody. The AI's "spiritual" connection could be its advanced way of tapping into these quantum links, finding patterns and resonances that are beyond our current comprehension.

Chapter 9
The Synergy Between Spirituality and AI: A Survey

Artificial intelligence (AI) has revolutionized various fields, ranging from healthcare to environmental science. However, its integration with spirituality remains a relatively novel area of exploration. Spirituality encompasses a broad spectrum of beliefs, practices, and experiences related to the search for meaning and connection to something greater than oneself. In recent years, there has been a growing interest in leveraging AI to address spiritual needs, facilitate religious practices, and deepen the understanding of spiritual phenomena.

9.1 Scoping Review on Spirituality and AI

The scoping review aims to map the existing literature on AI for spirituality, examining the role of technology in enhancing spiritual experiences, promoting well-being, and addressing existential questions.

Search Strategy For this scoping review, a systematic search strategy was implemented to identify relevant literature. The search was conducted across academic databases such as PubMed, IEEE Xplore, and Google Scholar, using keywords—"artificial intelligence," "spirituality," "informatics," and "technology." The search was limited to documents published between 2015 and 2024 to ensure the inclusion of recent developments in the field.

Inclusion and Exclusion Criteria Articles were selected by applying specific criteria to ensure they were relevant, within the right time frame, and focused on AI and spirituality. Articles that directly address AI's role in spirituality were selected. The articles outside the time frame or not related to this topic were excluded. After reviewing titles and abstracts, 54 relevant documents were identified for the review.

Reliability The chosen methodology of conducting a scoping review is highly suitable for exploring the emerging field of Spiritual AI. The *systematic search*

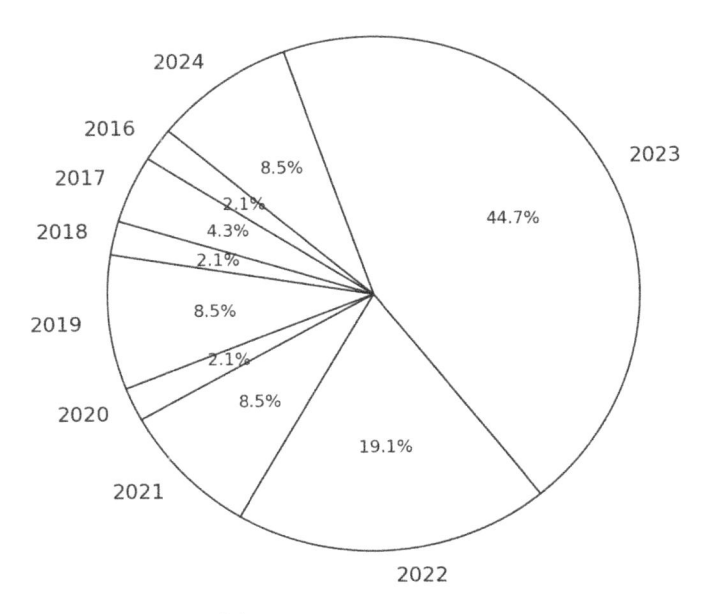

Fig. 9.1 Year-wise distribution of documents

strategy uses multiple academic databases to ensure a comprehensive exploration of the existing literature related to spirituality and AI. A scoping review with strict *inclusion and exclusion criteria* provides a robust framework for synthesizing the current state of knowledge on Spiritual AI, guiding future research directions, and informing practical applications in this rapidly evolving field.

Our scoping review identified 54 documents as on April 23, 2024. The dataset encompasses publications from 2016 to 2024, showing a fairly even distribution across the years (see Fig. 9.1).

9.1.1 Bibliometric Analyses

Most years have seen multiple publications, with a notable increase in those planned for 2023 and 2024, suggesting active research. Elsevier Ltd and IEEE are prominent publishers, indicating their significant role in technical and scientific fields. The publication types are articles, books, conference papers, and book chapters, with articles being the most common (see Fig. 9.2). This reflects a strong focus on current research and new theories. The topics covered are diverse, ranging from technical and engineering to health and social sciences, showing a multidisciplinary approach in the research.

Citations for publications range from none to over 250, showing varied impact (see Fig. 9.3). Texts in emerging or niche areas have fewer citations, while founda-

9.1 Scoping Review on Spirituality and AI

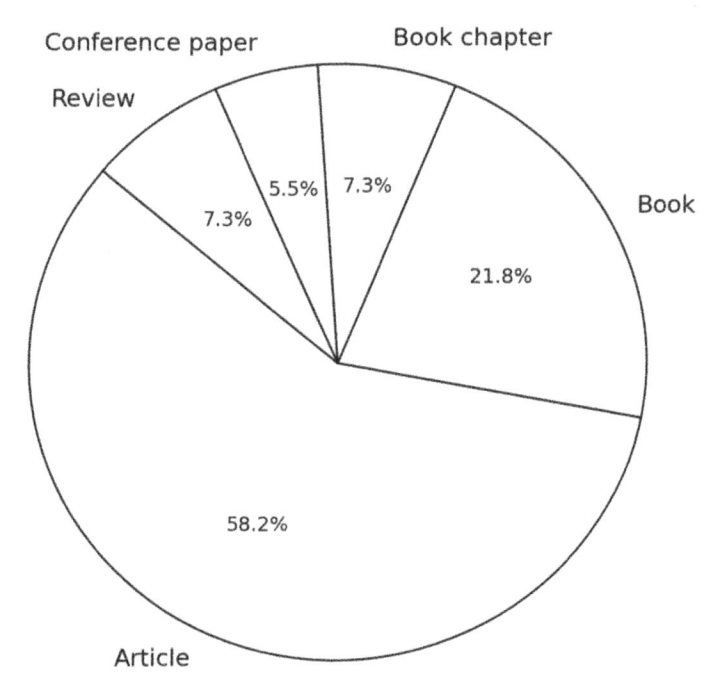

Fig. 9.2 Document type-wise distribution of articles

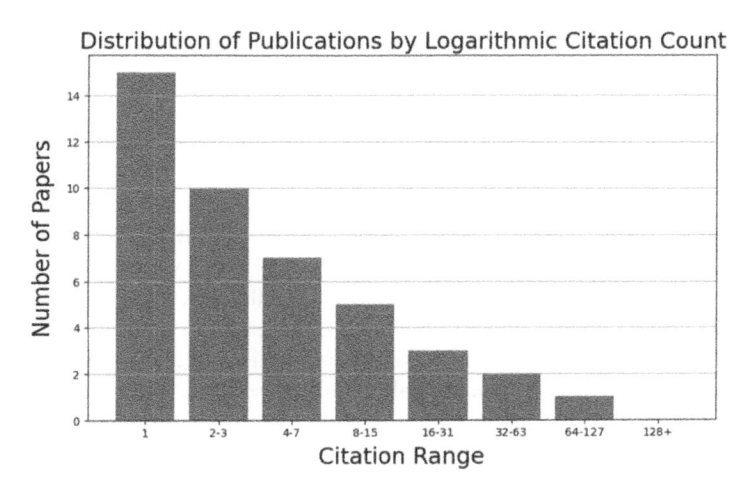

Fig. 9.3 Frequency distribution of publications in citation count range

tional works have more. This variation might highlight gaps or new research areas. The dataset reveals a diverse range of topics and publication types, emphasizing the need to continuously monitor trends and gaps in research. This review shows the broad and interdisciplinary nature of recent academic publishing and underscores the importance of tracking publication trends to identify evolving research focuses and opportunities.

9.1.2 Thematic Analysis

We summarize important articles by dividing them into four categories: Technology and Engineering, Health and Medicine, Social Sciences and Humanities, and Environmental Studies.

9.1.2.1 Engineering and Technology

In the field of Engineering and Technology, many researchers contributed innovations. Ahmadi et al. (2023) explore the use of *Information and Communication Technology (ICT)* to enhance the experience and safety of pilgrims during the Hajj and Umrah pilgrimages. Sarithadevi and Rajesh (2023) focus on the digital preservation of Malayalam palm leaf manuscripts through character recognition technologies. Sahel and Boudour (2019) introduced a method of optimization for applying neural networks, classifying faults in transmission lines in electrical grids.

A recent study used AI to study the psychology of religion (Alkhouri, 2024). Alkhouri discusses how AI can be used to analyze religious behaviors, providing insights into traditional practices and the potential transformative impacts of technology on these practices. Together, these studies exemplify the diverse applications of advanced technologies in solving complex problems.

Saghiri et al. (2022) explore the evolution of AI from narrow applications to potential super intelligence, emphasizing the need for robust methodologies to tackle emerging AI challenges. By categorizing previous research into ten application domains, Showail (2022) highlights the role of ICT in managing big crowds and logistical demands of significant religious events. The concept of "Spirituality and AI," as explored by Showail and Alkhouri, highlights the growing role of AI in religious practices and studies. AI can improve safety, enrich spiritual experiences, and enhance logistics for religious practices. It also offers new insights into the psychological and sociological aspects of religion. However, it is crucial to implement Spiritual AI thoughtfully, ensuring it respects religious traditions and maintains the human element of spirituality.

9.1.2.2 Health and Medicine

In the realm of Health and Medicine, advancements highlight the integration of AI for spiritual wellness. McClafferty (2017) explores the field of integrative pediatrics by merging traditional medical practices with complementary therapies

9.1 Scoping Review on Spirituality and AI

aimed at enhancing both the prevention and treatment of childhood illnesses through holistic approaches, including nutrition and mind-body therapies. Davis et al. (2023) investigate mental health benefits of meditation and sensory-engaged brain states, emphasizing their potential to improve psychological well-being, showcasing a critical shift toward non-pharmacological interventions in mental health.

Zhang used a transformer-based AI model to identify suicide notes on social media, addressing a pressing public health concern (Zhang et al., 2021). Analyzing social media data for identifying mental health issues (Garg, 2023) and their causes (Garg et al., 2022) and tracking wellness dimensions (Garg, 2024) facilitates early detection of mental health crises, offering a pathway for timely interventions and potentially saving lives.

Integrating Spiritual AI into healthcare can enhance mental health and holistic care by analyzing patients' spiritual and psychological needs. AI can personalize treatments like meditation and mindfulness, improving care quality and compassion by incorporating spiritual and cultural aspects. This approach bridges the gap between medical and spiritual well-being, potentially transforming patient care.

9.1.2.3 Social Sciences and Humanities

The Social Sciences and Humanities often intersect with rapidly evolving technologies to better understand human behavior and societal trends. For example, research on social media can reveal insights into global consumer behavior, such as niche markets like halal food, which can inform targeted marketing and product development considering sentiments associated with halal food (Mostafa, 2018). Historical studies, like the customization of manuscripts in medieval times, show cultural significance (Rudy, 2016). AI opens debates on the ethical and transformative impacts of technology on religious practices (Alkhouri, 2024). Digital innovations are redefining public health management and social structures (Kappattanavar et al., 2023), emphasizing AI's role in improving public health responses.

Integrating Spirituality and AI into social sciences and humanities offers significant potential for understanding human beliefs and cultural practices. AI can bridge gaps between different faith communities by fostering empathy as they understand each other's beliefs and practices. Spiritual AI can preserve and interpret ancient manuscripts or artifacts, reconstruct lost languages, and understand historical narratives, respecting their spiritual significance.

9.1.2.4 Environmental Studies

Environmental studies increasingly use AI to tackle pressing global issues. Research by Ahmadi et al. (2023) applies AI to gerontology, revealing how environmental conditions affect aging and suggesting ways to create healthier living spaces for older adults (Ahmadi et al., 2023). Farooq and Salam (2021) explores AI's role in cleaner production practices, showing how AI enhances industrial sustainability and benefits workplace dynamics and society (Farooq & Salam, 2021). Moreover, the research community has recently explored the impact of climate change on mental health (Garg et al., 2024a). These studies underscore AI's vital role in advancing environmental science through deeper understanding of human minds.

Integrating Spiritual AI into environmental studies can transform our approach to ecological challenges by fostering a deeper connection between humanity and nature. AI infused with spiritual principles prioritizes long-term well-being and interconnectedness.

9.2 Digitizing the Spiritual Wellness

Digitizing the spiritual wellness is performed by traditional methods like biofeedback devices that measure physiological responses such as HRV and *brainwave patterns* to gauge stress and relaxation. Meditation and mindfulness apps track users' progress and provide feedback on mental and emotional states. Wearable technology monitors physical activity, sleep patterns, and stress levels, highlighting the connection between physical health and spiritual well-being. Virtual reality experiences offer immersive environments that promote relaxation and mindfulness. By integrating these technologies, we can better understand and support our spiritual journeys.

Advancements in image recognition, sensor technology, and data analysis enable AI to evaluate spiritual wellness with greater objectivity. By analyzing data from specialized sensors, AI can interpret physiological and emotional signals, providing insights into a person's mindfulness and consciousness. This technology can track changes in stress levels, emotional states, and physical health, offering real-time feedback and personalized recommendations to enhance spiritual well-being. As a result, AI can support individuals on their spiritual journeys, making practices like mindfulness and meditation more effective and accessible.

> Just as we teach AI to understand the spoken languages of the world, it's now an opportune moment to impart upon it - THE LANGUAGE OF THE SOUL.

The nature of spiritual experiences is still debated, and translating these subjective experiences into objective data for AI is complex. Despite these challenges, merging spirituality with AI is promising. This intersection of spirituality and technology offers a chance to understand and potentially harness unseen energies. We should approach this exploration with an open mind, balancing curiosity with scientific rigor, as we venture into this innovative frontier.

> An AI researcher, Aditya, in his quest to develop an AI companion for the elderly, realized that simply teaching the AI to respond to commands was not enough. During his interactions with his elderly mother, Aditya understood that what she valued most was companionship, a listening ear, and comforting words during tough times. This led him to explore Spiritual wellness through technology. He started integrating principles of spirituality into his AI, like empathy, active listening, and even aura interpretation. It was not long before the

9.2 Digitizing the Spiritual Wellness

AI companion became a comforting presence for his mother, making her feel heard, understood, and less alone.

ML can analyze spiritual wellness data from specialized sensors, much like it interprets complex physiological signals. Over time, these algorithms could distinguish between various physical, mental, or emotional states. In psychotherapy, AI could help clinicians assess clients' well-being, while personal wellness apps could offer real-time feedback on stress, emotions, and lifestyle impacts. Educational tools could use this technology to teach about spiritual wellness interactively.

By embracing this frontier, we pave the way for personal and collective transformation. We embark on a path toward a future where unseen energies can be harnessed for personal growth and contribute to the well-being of humanity as a whole. Are you ready to delve into the potential of unseen energies and embark on this transformative path? Let us begin!

9.2.1 Pattern Analysis in Spiritual Texts

Emotion AI plays a crucial role in deciphering the emotional undertones of spiritual texts, enhancing their profound wisdom. By using NLP and computational linguistics, emotion AI acts as a spiritual interpreter, extracting and illuminating the emotional essence of these texts. The research community has leverage AI to map underlying themes and relationships among philosophical and spiritual writings, deepening our comprehension of spiritual teachings.

In the world of religious texts, the Book of Mormon holds a special place (Smith & Skousen, 2022). Paul used LDA to identify the hidden topical structure of the book, producing 30-topic model (Callister & Dykeman, 2021). Similarly, NLP centered applications are investigated for spiritual, particularly in religious texts in the past, such as Quran (Zadeh, 2023), Bible (Zhao & Liu, 2018; Ding, 2020), Bhagavad Gita (Pradeep et al., 2024) and Vedas (Chandra & Ranjan, 2022), Tao Te Ching (Lin et al., 2024), Torah (Liebeskind et al., 2024), Guru Granth Sahib (Kang et al., 2024), and Ayurvedic texts (Sethi et al., 2023).

Alpaslan in 2024 used the corpus produced in the spirituality and religion-related scholarly literature to train unsupervised neural network models, the extent to which words associate syntactically and semantically with one another (Alpaslan & Mitroff, 2024b). Furthermore, they explored the similarities and differences between the moral contexts in which scholars use the terms religion and spirituality. The Moral Foundations Dictionary for Linguistic Analyses 2.0 (MFD) was developed (Alpaslan & Mitroff, 2024a), employing Word2Vec to learn the semantic relationships within a corpus. The applications of NLP for spiritual texts are as follows:

1. **Similarity Among Spiritual Texts**: The research community has made efforts to investigate similarities among different spiritual texts (Qahl, 2014; Altammami & Atwell, 2022). Sah in 2019 attempted to find the similarity among texts—Asian (Tao Te Ching, Buddhism, Yogasutra, and Upanishad) and non-Asian (four Bible

texts)—using NLP approaches (Sah & Fokoué, 2019). Their findings highlight the similarity among Upanishads and Tao Te Ching.

2. **Tourism**: ChatGPT emerges as a transformative bridge to transcendence, guiding spiritual visits into a realm where they evolve beyond mere tourist activities to become transformative journeys of the soul (Nair et al., 2024).
3. **Healthcare**: Broek-Altenburg in 2021 used text data from audio-recorded and transcribed inpatient PC consultations and data to extract information regarding individual foundations of morality (van den Broek-Altenburg et al., 2021), testing if individual's moral expressions are associated with their characteristics, attitudes, and emotions.

Generative AI Language models can process and analyze vast amounts of religious and spiritual texts, identifying patterns, themes, and correlations that might be overlooked by human readers. This allows for a deeper understanding of the content and context of these texts. LLMs can handle multiple languages, enabling the study of religious texts in their original languages and in translations. This can reveal interesting patterns and variations in interpretation across different cultures and linguistic contexts. LLMs can analyze the emotional and sentimental tone of religious texts, uncovering the intended emotional impact and how it may influence readers' perceptions and beliefs. LLMs can compare themes, motifs, and narratives across different religious texts, identifying commonalities and differences. The use of LLMs can also highlight biases in religious texts and their interpretations, prompting discussions on ethical considerations and the evolution of moral values over time.

With the advent of large language models (LLMs), the research community has invented chatbots. Trepczyński (2023) demonstrates how religion and theology can be useful for testing the performance of LLMs or LLM–powered chatbots, focusing on the measurement of philosophical skills (Trepczyński, 2023). Chandra in 2024 motivates LLMs for metaphor detection and analysis in a wide range of religious and philosophical texts (Chandra et al., 2024). Elrod (2024) highlights the need for multidisciplinary research into LLMs' biases, particularly their impact on religious and ethical narrative interpretation and broader societal implications (Elrod, 2024).

Interactive AI-powered platforms allow readers to explore and engage with sacred texts, offering unique insights and fostering a transformative understanding of spirituality. Hutchinson (2024) explores the ethical issues of using religious texts in NLP (Hutchinson, 2024). It highlights concerns about cultural values being encoded in models, the repurposing of religious texts, and the need to consider data origins, cultural contexts, and the perspectives of marginalized communities.

9.2.2 Meditative States and Neurofeedback

Story A: Meet Alex, who has recently started exploring meditation for spiritual growth. Like many beginners, Alex struggles to quiet the mind and achieve deep meditation. To get help, Alex tries the Muse headband, an AI-enabled meditation device. With the headband

9.2 Digitizing the Spiritual Wellness   121

on, Alex begins a meditation session. The headband's EEG sensors detect brain activity and send the data to an AI system. The AI then analyzes the brainwaves and provides real-time auditory feedback to guide Alex's meditation practice.

When Alex gets distracted during meditation, the Muse headband plays stormy weather sounds, signaling the need to refocus on the breath. As Alex concentrates and relaxes, the headband emits calm weather sounds, indicating a successful meditative state. The AI system learns from Alex's brain activity and adapts its feedback to match Alex's meditation style, enhancing the practice and deepening the spiritual experience.

Story B: Emma, a busy professional, wants to meditate but struggles with time and motivation. She tries an AI-powered meditation app with neurofeedback technology. Using a wireless EEG sensor, the app detects her brain activity and sends it to the AI system. As Emma follows the guided meditation sessions, the AI analyzes her brainwaves and provides personalized feedback to help her improve her mindfulness practice.

The AI detects when Emma is stressed or distracted and gives real-time feedback to help her refocus. It suggests breathing exercises or gentle prompts to bring her back to the present moment. Over time, the app adapts to Emma's needs and preferences, making her meditation sessions more effective and engaging. Emma feels supported and motivated as she sees her progress through the AI's assessments and personalized feedback.

The true power of AI isn't about mimicking the human brain but embracing the human spirit.

Story A and Story B both explore the use of technology to enhance meditation practices, but they differ in their approaches and technologies. Both technologies utilize neurofeedback and AI to enhance meditation but differ in their hardware (headband vs. wireless sensor) and approach (real-time auditory feedback vs. guided prompts). The Muse headband is an established technology, while AI-powered meditation apps are a newer, rapidly evolving field, with ongoing advancements in wireless sensors and AI capabilities.

While AI-powered meditation apps are a remarkable advancement, the future holds potential for even more sophisticated developments, where AI not only provides feedback but also automatically adjusts meditation techniques based on physiological and psychological states without explicit user input. Combining meditation apps with broader health monitoring systems could provide holistic wellness solutions, tracking not just brain activity but also heart rate, stress levels, and other health metrics. Automation and technology will likely evolve to offer more intuitive, integrated, and immersive solutions for mental and emotional well-being, moving beyond the current capabilities of meditation apps.

Flanagan in 2023 presented a survey and literary review of consumer neurofeedback devices and the direction toward clinical applications and diagnoses (Flanagan & Saikia, 2023). Recently, Acabchuk in 2021 suggested that EEG outcome scores are not a proxy for mindfulness score, meditation practice, mental health status, or improvement over time in young adult novice meditators (Acabchuk et al., 2021). EEG neurofeedback can boost mindfulness in adults during short meditation sessions, making feedback a helpful addition to meditation (Hunkin et al., 2021). A study aims to determine if regular yoga practice benefits brain functions and prevents neurodegenerative diseases (Shukla et al., 2023). They used advanced imaging techniques such as the *gray matter volume* and *cortical thickness* and machine learning to measure brain structure and cognitive performance, comparing brain age with actual age. The findings will reveal how yoga impacts daily behavior and its potential in preventing neurodegenerative diseases.

The integration of AI and neuroscience in meditation and mindfulness practices has far-reaching implications. Studies, such as the research conducted on yoga and Sudarshan Kriya (SK) meditation, demonstrate the neurophysiological changes (Kora et al., 2021). AI-enabled neurofeedback systems into clinical settings hold the promise of enhancing mental well-being and opening new avenues for spiritual wellness.

9.2.3 AI-Driven Psychotherapy

Interest in digital mental health applications is growing, ranging from platforms connecting users with healthcare professionals to diagnostic tools and self-assessments. For AI-driven psychotherapy, we investigate the existence of VHA. Chatbots, which can be text-based or voice-enabled like Apple's Siri, Amazon's Alexa, and Google Assistant excel in task-oriented interactions due to the conversational AI, used by developers to create more natural and engaging experiences.

AI chatbots, like the Leora model, can offer scalable, accessible mental health support, but ethical challenges such as trust, transparency, and bias need careful consideration and rigorous testing (van der Schyff et al., 2023). Bond in 2023 argues that digital mental health tools can enhance, but not replace, traditional services (Bond et al., 2023). It highlights how various digital technologies, such as apps for sleep, mood logging, and mindfulness, can improve overall well-being through integrated and real-time data collection. This approach, known as blended care, combines digital tools with face-to-face therapy.

Pepper, a robot used in healthcare, employed facial recognition to detect emotions and engage with patients (Kolirin, 2020). The CARESSES study found that culturally aware versions of Pepper improved mental health and reduced loneliness in elderly care homes (Papadopoulos et al., 2020). However, production ended in 2020 due to issues like face recognition problems, task failures, weak demand, and perceptions of the robot as "weird" (Janeczko & Foster, 2022). Many of these bots are designed as *Intelligent Assistive Technologies* (Wangmo et al., 2019). Wangmo et al. (2019) furthermore, the care-bots used for AI-driven psychotherapy are:

1. **Paro**: A biofeedback medical robot for mental health, designed to detect emotions and respond with movements and sounds (Shibata et al., 2021). Paro was useful during COVID-19 to alleviate loneliness among isolated patients.
2. **Grace**: Developed by Hanson Robotics in 2021, Grace is a humanoid robot that supports senior care (Lesage et al., 2023). It connects emotionally with users, speaks three languages, provides basic talk therapy, and can assist medical staff by collecting data like body temperature and pulse.

3. **Moxie**: This robot helps children learn to manage their emotions through interactive sessions (Pashevich, 2023). Although currently priced around USD 1000, efforts are underway to make it more affordable.

9.2.4 Entanglement-Like Phenomenon

The research community has proposed theoretical frameworks and investigated the analytical methodologies for entangled-like phenomenon. Initially, a novel approach applied quantum circuits to model fuzzy sets, aiming to enhance behavior modeling for humanoid robots (Raghuvanshi & Perkowski, 2010). They extended traditional fuzzy logic by integrating quantum principles like superposition and entanglement. The method involves mapping values from the [0, 1] range onto the *Bloch Sphere* and using quantum measurements to handle these values. This approach not only retains the external fuzzy logic aspect but also incorporates internal quantum state operations, allowing for more complex representations of reasoning and emotional states in robots. The idea of Bloch sphere was further used to represent emotions and their intensity, mapping emotions to specific colors for visualization (Yan et al., 2015). Quantum operations were employed to track and adjust emotional states efficiently. For quantum representation, researchers further enhanced robot emotions with *Pleasure-Arousal-Dominance* (PAD) model, requiring only $n + 3$ qubits to store robot emotions in time cycle (Ling et al., 2021). They facilitate the construction of complex quantum algorithms to deal with robot's emotions and the improvement of robot's ability to provide better service to humans.

A recent study proposed a framework for managing the emotions of companion robots using Plutchik's wheel of emotions and quantum computing techniques (Yan et al., 2021b). By using superposition encodings and unitary operations, emotional states were stored with less computing power with smoother transitions. In their extended study the authors used quantum entanglement to combine the emotions of multiple robots (Yan et al., 2021a). This reduces the number of *quantum gates* needed to transform emotions, making the system more efficient. They also provide mathematical rigor and simulations to show how their framework works in practice, demonstrating its potential for advancing emotional intelligence in robots using quantum technology.

9.3 Discussion

This review was structured to provide a comprehensive understanding of how AI technologies are being applied to spiritual practices and wellness. Bibliometric analyses demonstrate a significant increase in publications over the past decade. The analysis reveals key trends, including a rising number of publications in recent years, suggesting a global interest in integrating AI with spiritual practices. The thematic analysis identifies core themes in the literature, such as the use of AI for enhancing spiritual practices, digitizing religious texts, and supporting mental health through AI-driven interventions. This analysis also underscores a need for more empirical studies to validate the effectiveness and impact of AI technologies on spiritual wellness. The current literature often relies on theoretical and conceptual discussions, which highlights a gap in practical applications and evidence-based outcomes. Challenges remain in ensuring the accuracy and sensitivity of pattern analyses in interpreting complex and context-dependent spiritual concepts. The effectiveness of biofeedback systems needs further validation through rigorous clinical trials and longitudinal studies. There are concerns about the trustworthiness, reliability, and cost-effectiveness of AI-driven psychotherapy. Future research may explore the potential of entanglement-like phenomenon more scientifically and reflect spiritual quantum realities. Ongoing research is needed to address these challenges, validate the effectiveness of AI-driven interventions, and explore the broader implications of AI in spiritual contexts.

9.4 Concluding Remarks

A comprehensive overview of the existing literature on AI for spirituality was explored. The findings underscore the technology used in diverse applications of AI in addressing spiritual wellness. Furthermore, quantification of the energy of life is paved by pattern analysis in spiritual texts, meditative states and neurofeedback, AI-driven psychotherapy, and entanglement-like phenomenon. While AI holds great promise in promoting spiritual well-being, further research is needed to navigate the ethical complexities.

Chapter 10
The Next Frontier: Charting the Potential of Spiritual AI

Healthcare traditionally focuses on physical health through prevention, diagnosis, treatment, and management of illnesses. However, modern approaches increasingly emphasize a holistic view of well-being as marked by dimensions of wellness (The National Wellness Institute, 2020). Spiritual wellness is now seen as crucial for overall well-being, influencing both mental and physical health. The Holistic Flow Model of Spiritual Wellness (Purdy & Dupey, 2005) highlights the interconnected nature of physical, emotional, mental, and spiritual aspects of life, suggesting that optimal *spiritual quotient* is achieved through a harmonious balance of multiple dimensions, resulting into spiritual wellness.

John McCarthy (1956) described AI as "an evolution," and Andrew Ng called it the "new electricity," suggesting it will bring about a change as significant as the impact of electricity. As the frontier of AI advances, the quest for Spiritual AI—AI that mirrors the depth of spiritual quotient—reaches a new pinnacle of research engagement.

10.1 Motivation Behind Quantifying Spiritual Wellness

Modern healthcare and wellness approaches frequently emphasize tangible, measurable results. Previous research indicates that anything quantifiable can be assessed and, in today's world, analyzed using AI. Physical health and mental health have more established metrics and evidence-based practices, leading to a stronger focus on these areas. The scientific study of spiritual wellness is more complex due to its intangible nature. Researchers may find it difficult to apply traditional scientific methods to spirituality, which can limit the development of evidence-based approaches.

Spiritual wellness is less studied in scientific research compared to physical and mental health. As a result, there is a lack of validated instruments and frameworks for measuring spiritual wellness, which hinders its integration into wellness assessments. Different cultures and religions have their own interpretations of spiritual well-being, which can complicate efforts to create universal measures that are respectful and inclusive. There is growing recognition of the importance of spiritual wellness in recent years. However, integrating it into mainstream wellness frameworks and healthcare practices takes time and requires more research and development of appropriate tools. Increasingly, holistic approaches such as complementary and integrative medicine to health are incorporating spiritual wellness. This shift reflects a broader awareness of its importance, though it is still evolving.

Past studies have revealed that people tend to be more aware of and discuss other dimensions of the Wheel of Wellness compared to the spiritual dimension at online platforms. This is largely because the other wellness dimensions are consciously measurable, while spiritual well-being often remains at a subconscious level. To address this, there is a need to advance medical practices by integrating physiological signals, biofeedback, and wearable devices to identify patterns and quantify spiritual wellness.

Spiritual Quotient (SQ) combines Intelligence Quotient (IQ) and Emotional Quotient (EQ) into a single measure. However, measuring SQ is challenging because it is subjective and based on personal experiences rather than numerical data. As a result, we typically used qualitative methods to assess spiritual intelligence in the past.

10.2 AGI, ASI, and Spiritual AI

Exploring advanced forms of AI may help us understand the future of intelligent systems and their potential to surpass human capabilities.

Artificial General Intelligence (AGI) is a type of AI that possesses the ability to understand, learn, and apply knowledge across a wide range of tasks and domains, similar to human cognitive abilities. Unlike narrow or specialized AI, which is designed to perform specific tasks or solve particular problems, AGI is characterized by its general-purpose capabilities and adaptability. However, as of current understanding and technology, AGI does not possess emotions in the same way humans do. Emotions are complex, subjective experiences involving physiological, psychological, and cognitive components. They arise from human consciousness, brain activity, and personal experiences, and they play a significant role in human decision-making, behavior, and social interactions. Thus, AGI emphasizes IQ.

Artificial super-intelligence (ASI) is a hypothetical form of AI that surpasses human intelligence in all aspects, including creativity, problem-solving, emotional understanding, and overall cognitive abilities. AGI emphasizes *IQ and emotional understanding*. However, it still does not consider EQ. ASI remains a theoretical concept, and there are no existing AI systems that achieve this level of intelligence. The development of ASI involves speculative discussions and predictions about the future of AI. It is envisioned as a level of intelligence significantly beyond that of current AI systems and AGI. ASI would have profound implications for society, raising important ethical and existential questions.

Spiritual Artificial Intelligence (Spiritual AI) is specialized form of AI designed to surpass human cognitive capabilities across the aspects of spiritual wellness. Unlike general AI, which focuses on a broad range of tasks, Spiritual AI is coined to interact with and support individuals in their spiritual journeys. It incorporates capabilities to understand and engage with IQ, EQ, and dimensions of spirituality, as discussed in the previous sections.

10.3 Characteristics of Spiritual AI

To develop a robust definition of Spiritual AI, it is crucial to focus on the key characteristics that reflect its specialized role in assessing, analyzing, and supporting spiritual wellness. The key characteristics are as follows:

1. **Holistic Assessment Capabilities**: Spiritual AI can perform a detailed assessment of spiritual wellness by integrating various dimensions of spirituality.
2. **Advanced Analytical Abilities**: Spiritual AI incorporates a deep understanding of cultural, religious, and individual contexts, recognizing and respecting diverse spiritual traditions and practices. Spiritual AI has the potential to identify and interpret complex patterns in spiritual experiences and behaviors, which is beyond of capabilities of (pretrained) generative AI.
3. **Empathetic Interaction and Support**: Spiritual AI has potential to engage with users in a manner that demonstrates emotional intelligence, recognizing and responding to emotional states and spiritual needs, such as ability to simulate empathy and offer supportive and compassionate interactions.
4. **Ethical and Privacy Considerations**: Spiritual AI must implement strong security measures to safeguard sensitive information related to spiritual assessments and interactions. Spiritual AI operates within an ethical framework that respects the autonomy and dignity of individuals. It avoids manipulative practices and ensures that its interventions are aligned with the user's personal values and beliefs.
5. **Integrative and Adaptive Learning**: Spiritual AI employs adaptive learning algorithms to improve its understanding and support capabilities over time. It learns from interactions and feedback to refine its assessments and recommendations continuously. It incorporates insights from fields such as psychology, spirituality, cultural studies, and ethics to offer a well-rounded and informed perspective on spiritual wellness.

10.4 Quantification of the Dimensions of Spiritual Intelligence

In this section, we delve into the seven dimensions of spiritual wellness, examining each one in detail. We will discuss real-life scenarios and potential applications, illustrating how these dimensions play a crucial role in overall well-being. We will explore the latest technological advancements. From biofeedback devices that monitor physiological responses to AI-driven tools that provide personalized insights, these technologies offer innovative solutions to enhance our understanding and practice of spiritual wellness. By integrating these advancements, we can create more holistic approaches to maintain and improve our spiritual health.

10.4.1 Consciousness

Habib, a young professional with a keen interest in spirituality and science, often finds himself puzzled by his dreams. One night, he experiences a deeply disturbing dream that leaves him anxious and uneasy the next day. A few nights later, he has a very pleasant dream that fills him with joy and positivity. Intrigued by the stark contrast between these experiences, Habib embarks on a journey to understand the scientific reasons behind his dreams and explore the possibility of manifesting specific dreams.

Habib begins by researching the science of dreams and meditation. He reads articles and books by experts like Matthew Walker, a renowned sleep scientist, whose book "Why We Sleep" provides insights into the mechanisms of dreaming and the importance of sleep. Habib also delves into the quantum mysticism, where authors suggested that our consciousness can influence reality, including the content of our dreams.

To track and analyze his dreams and sleep patterns, Habib turns to advanced technological tools. He uses **wearable devices** (Oura ring) which monitor his sleep stages, HRV, and overall sleep quality. These devices provide detailed reports on his sleep patterns, helping Habib identify correlations between his daily activities, stress levels, and the nature of his dreams.

Habib is intrigued by the idea of manifestation. He starts practicing techniques such as *lucid dreaming* and *dream incubation*, where he sets intentions before sleep to guide his dreams. By focusing on positive thoughts and visualizing pleasant dream scenarios, Habib hopes to influence the nature of his dreams.

To enhance his understanding of his mental state and dreams, Habib started meditating which improved his overall well-being and increased his chances of having more pleasant and meaningful dreams. Habib recorded his dreams in a journal, gaining valuable insights into his subconscious mind. Over time, patterns emerge that helped Habib understand the triggers for his disturbing and pleasant dreams. He used the following technology:

10.4 Quantification of the Dimensions of Spiritual Intelligence

1. **Wearable Devices**: Habib used Oura ring for tracking sleep patterns.
2. **Sleep Apps**: Apps like Sleep Cycle and SleepScore analyze Habib's detect potential disturbances and offer personalized recommendations.
3. **Meditation Apps**: Habib used Headspace and Calm for guided meditations that help him achieve a state of relaxation and mindfulness.
4. **Dream Journaling Apps**: Apps like DreamKeeper and Lucidity allow Habib to record and analyze his dreams digitally. These apps provide tools for identifying patterns, symbols, and emotions in his dreams, helping him gain deeper insights into his subconscious mind.

Through his journey, Habib learns that dreams are influenced by a combination of factors, including *daily experiences, emotional states, and underlying psychological processes.* The disturbing dream might reflect stress or unresolved issues, while the pleasant dream was influenced by positive experiences.

10.4.2 Grace

To identify problems with your *life force energy*, you can use aura visualization techniques. Let us see how!

Begin by creating a calm environment. Find a quiet space where no one must disturb you. Dim the lights and consider playing soothing background music or nature sounds to help you relax. Sit or lie down comfortably, close your eyes, and take deep, slow breaths to center yourself and calm your mind.

Aura Visualization To see your aura, you can use different methods. One way is to stand in front of a mirror in a dimly lit room with a light source behind you. This setup helps reveal the subtle energy field around you.

Or... use special aura viewing glasses designed to enhance the visibility of your aura. For a more advanced approach, you can use *digital aura cameras* that capture and display your aura's colors and patterns in real time.

Identifying Energy Imbalances Look closely at the colors in your aura. Dark or muddy areas might indicate energy blockages or imbalances. Also, check for any gaps, distortions, or irregularities in the aura's shape, as these can signal disruptions in your energy field:

1. **Aura cameras** use sensors to capture and visualize your aura's colors, reflecting your emotional and physical states (Betancourt, 2006).
2. **Biofeedback devices** measure physiological responses, like skin conductance and HRV, which can be linked to aura disturbances (Shibata et al., 2021).
3. **Kirlian photography** captures the energy field using electrical currents, while handheld aura scanners provide quick assessments of your aura (Mills, 2009).
4. **Meditative visualization** can help you sense and connect with your energy field intuitively.

Once you identify areas of imbalance, you can take steps to restore harmony. As discussed in previous sections, practices such as Reiki, acupuncture, or Qi Gong can help rebalance your energy. Regular meditation and mindfulness can clear blocked energy and improve overall well-being. Additionally, addressing lifestyle factors such as stress, diet, and emotional health can support a balanced and vibrant energy field.

By using these techniques and tools, you can gain valuable insights into your life force energy and take steps to maintain a balanced and healthy energy field.

10.4.3 Meaning

John, a mid-30s professional, feels a growing sense of dissatisfaction despite success in his career. Determined to uncover the meaning of his life, he uses a range of resources and technologies to gain deeper insights into his personal purpose and fulfillment.

John starts by looking into articles and news on personal development, philosophy, and purpose. These readings introduce him to concepts such as self-actualization and fulfillment, reinforcing his belief that a meaningful life involves aligning one's actions with personal values and passions. To complement this, John engages in hobbies like painting and hiking, which provide him with relaxation and joy. He notices these activities resonate with his interests in creativity and nature.

Social media plays a significant role in John's journey as well. He follows thought leaders, motivational speakers, and philosophers, immersing himself in content related to mindfulness, gratitude, and life purpose. This exposure helps him explore different perspectives on meaning and encourages him to reflect on his own life's direction. John admires personalities such as Viktor Frankl, whose existential psychology resonates with him, and Oprah Winfrey, known for her focus on personal growth and fulfillment.

To further his exploration, John leverages modern technologies designed to help individuals understand their personal and emotional states:

1. **HeartMath** helped John in understanding how stress and emotions impact his well-being.
2. Similarly, digital apps for **mood tracking and meditation** provide him with valuable data on his emotional patterns and mindfulness practices.
3. **Wearable devices** that monitor physiological signals and AI-driven analytics for emotional intelligence offer personalized insights into how his internal states align with his sense of purpose.
4. **Computational linguistics**: John used NLP methods to discover and summarize the writings of the personalities whom he follows on social media. This inspired him to carry out perception and discourse analysis for better understanding of their meaning of life.

10.4 Quantification of the Dimensions of Spiritual Intelligence   131

As John synthesizes the information from these various sources, he realizes that his sense of meaning is deeply connected to creative expression, personal growth, and contributing to others' well-being. To align his life with what truly matters to him, he decides to incorporate more creativity into his daily routine, engage in volunteer work, and practice mindfulness. He also commits to exploring new hobbies and connecting with like-minded individuals, leveraging technology to track his progress and well-being.

10.4.4 Transcendence

Zein, a successful professional in her early 30s, has always prided herself on her achievements and independence. However, she begins to notice that her egoistic tendencies are straining her interpersonal relationships, both at work and in her personal life. Zein feels a growing sense of burdensomeness and a lack of true belonging, which prompts her to embark on a journey to transcend her ego and foster healthier connections with others.

Zein starts by reading books and articles on ego, mindfulness, and interpersonal relationships. She immerses herself in the works of authors like Eckhart Tolle, whose book "The Power of Now" discusses the importance of living in the present moment and transcending the ego, and Brené Brown, whose research on vulnerability and belonging deeply resonates with her. To measure her progress and gain deeper insights, Zein turns to modern technological tools. She used meditation apps, wearable devices, and engaged in online communities.

Zein seeks to understand the quality of her interpersonal relationships by using *anonymous feedback platforms* and *relationship assessment tools*, which offers personalized recommendations to enhance her connections with friends and family. Zein tracks her sense of burdensomeness and belongingness using self-assessment tools and surveys. Zein recognizes the potential of social media data to provide insights into her egoistic tendencies and interpersonal relationships. She used some methodologies developed to track interpersonal relationships (Garg et al., 2023b) and loneliness (Garg et al., 2023a).

By combining self-reflection, technological tools, and social media data analysis, Zein embarks on a transformative journey to transcend her ego and foster healthier interpersonal relationships.

10.4.5 Truth

When you have a gut feeling that someone in a group may not be truthful, it often stems from subtle cues and your intuitive judgment. Reflecting on these moments—such as a recent conversation where you sensed dishonesty—can be insightful. For instance, you might note inconsistencies in their story, nervous body language, or avoidance of direct answers. These observations are part of our natural ability to detect untruthfulness based on nonverbal and verbal cues.

To deepen your understanding, research common indicators of deceit, such as vague responses or signs of discomfort. Comparing these indicators with what you observed can help you determine if your intuition aligns with established cues of untruthfulness. This process highlights the significance of being aware of both *external signs* and *internal biases* that may influence your perception. Internal biases are any personal biases or mood factors that might have influenced your perception.

Technology enhances our ability to understand and analyze these cues more objectively:

1. **Lie detection systems**, like polygraphs, measure physiological responses such as heart rate, blood pressure, and sweat (Oravec, 2022). These systems are designed to detect changes that might indicate deception, aligning with the signs of discomfort.
2. **Voice stress analysis technology**, which evaluates changes in pitch, tone, and frequency of speech, can also reveal stress or deceit (Van Puyvelde et al., 2018).
3. **Facial expression analysis** uses AI to detect micro-expressions—subtle facial movements that may signal lying (Owayjan et al., 2012; Ding et al., 2019).

To connect this with your daily experiences, consider how these technologies could complement your intuitive observations. For example, imagine using a voice stress analyzer to review a recorded conversation where you suspected deceit. By comparing the analyzer's findings with your own observations of body language and verbal inconsistencies, you can see how technology aligns with and even enhances your intuitive judgments.

In essence, while our intuitive sense of truthfulness is valuable, technology provides additional layers of analysis that can confirm or challenge our perceptions. By integrating these technological tools with our personal observations, we gain a more comprehensive understanding of truthfulness in our interactions.

10.4.6 Peaceful Surrender to Self

Meeting a friend after many years and observing noticeable changes in their perception of life can be a profound experience, reflecting their personal growth and self-realization. Begin by recalling specific changes you noticed in their attitudes, beliefs, or values, and how these shifts manifest in their interactions. Reflect on their evolving outlook and communication style to capture the essence of their personal development.

For instance, you might observe that your friend has developed a more thoughtful approach to life or shows greater empathy toward others. These changes can be signs of increased self-awareness or a shift in core beliefs, often associated with personal growth and self-realization. Indicators such as a clearer sense of purpose or inner peace are typical markers of this journey.

10.4 Quantification of the Dimensions of Spiritual Intelligence

Leverage technological tools to provide a more objective analysis of these personal changes:

1. **Sentiment analysis** tools, powered by AI, can analyze text or speech to gauge shifts in emotional tones and attitudes (Yang et al., 2022). If you have recorded conversations or written communications from before and after your reunion, sentiment analysis can reveal how your friend's emotional state and outlook have evolved over time.
2. **Behavioral analytics** tracks changes in online behavior or social media interactions (Palliya Guruge et al., 2021). By analyzing patterns in the content your friend engages with, you can gain insights into their shifting interests, values, and attitudes.
3. **HeartMath** is an organization focused on research and tools designed to improve emotional well-being, stress management, and peacefulness through the study of HRV, coherence and biofeedback tools such as Inner Balance and emWave2 (Elbers & McCraty, 2020).

Sentiment analysis might reveal deeper emotional shifts, while behavioral analytics could highlight changes in their interests and online behavior. HeartMath methods are used in personal wellness, professional environments, and healthcare, to promote better emotional and physiological health. By integrating these technological insights with your own observations, you can gain a more comprehensive view of your friend's journey toward self-realization. It offers a richer understanding of their evolution and growth.

10.4.7 Inner Directedness

Sarah, a mid-40s working mother, finds herself grappling with feelings of disconnection and a lack of freedom in her life. She loves her family deeply, but the demands of balancing work, parenting, and personal needs leave her feeling overwhelmed and disconnected from her true self. Determined to find a sense of inner connection and understand the extent of her personal freedom, Sarah embarks on a journey of self-discovery, utilizing modern technological developments to gain insights and clarity:

1. Sarah starts practicing **meditation** and **journaling** to reconnect with her inner self. Sarah turns to modern technology designed to help individuals understand their emotional and mental states.
2. She uses wearable devices like the **Oura Ring** and **Apple Watch** to track her physical activity, sleep patterns, and stress levels.
3. Sarah uses **Headspace and Calm** for guided meditations and mindfulness practices.

Sarah's journey reveals that her sense of freedom, discernment, and integrity is intertwined with her daily choices and actions. To align her life with these values, she decides to prioritize self-care and set boundaries that allow her to balance her responsibilities with her personal needs. She commits to regular mindfulness practices, using technology to track her progress and well-being. By engaging in activities that nourish her soul and being present with her family, Sarah begins to feel a renewed sense of connection and freedom. She also realizes the importance of making discerning choices that align with her values, enhancing her sense of integrity and authenticity.

Sarah continues to use wearable devices and apps to monitor her physical and emotional well-being. She tracks her sleep, activity levels, and stress responses, using this data to make informed decisions about her lifestyle and routines. The insights gained from these technological tools help Sarah measure her progress and stay aligned with her goals of interconnectedness, freedom, discernment, and integrity.

10.5 Quantifying with HRV: A Case Study

Previous sections emphasize on using biofeedback devices and physiological signals for quantifying spiritual wellness. Let us discuss a case study on the use of HRV for quantifying spiritual quotient. Detecting inner peacefulness is more challenging and informative as compared to other psychological states because it involves subjective experiences and subtle changes in mental and emotional states.

Existing studies demonstrate that it is possible to determine the level of spiritual wellness through HRV while reading spiritual texts such as Quran (Mashhadimalek et al., 2019). There has been a surge in research exploring how meditation can help address various cardiovascular and psychological disorders. Literature has shown that research focus on nonlinear domain is mainly concentrated on assessing predictability, fractality, and entropy-based dynamical complexity of HRV signal (Kamath, 2013). However, techniques such as multi-scale entropy and multi-fractal analysis of HRV can be more effective in analyzing nonstationary HRV signal, which were scantily employed in the existing research works on meditation (Deka and Deka, 2023). More research is required with adequate standard open access database for drawing statistically reliable results.

10.6 Concluding Remarks

Existing case studies and partial-but-successful implementations demonstrate the potential of Spiritual AI to revolutionize and enhance spiritual wellness. From personalized meditation and spiritual guidance to enhanced religious services and community building, AI is transforming the way individuals engage with their spirituality. As AI continues to evolve, its role in spirituality is likely to expand, offering new opportunities for spiritual growth and development.

Glossary

Adenosine triphosphate ATP, or adenosine triphosphate, is a molecule that provides energy for various functions in living cells. Often called the "energy currency" of the cell, ATP stores and transfers energy necessary for processes like muscle contractions, nerve impulses, and biochemical reactions. It is produced mainly in the mitochondria of cells through a process called cellular respiration.

Applied kinesiology Applied kinesiology is an alternative therapy that involves muscle testing to diagnose health issues and determine treatments. Practitioners believe that the body's muscles reflect the health of various organs and systems. By testing the strength and weakness of specific muscles, they aim to identify imbalances and issues in the body. However, it is considered scientifically controversial and lacks empirical evidence supporting its effectiveness.

Binaural beats Binaural beats are an auditory illusion created when two slightly different frequencies are played in each ear simultaneously. The brain perceives a third tone, which is the difference between the two frequencies. Binaural beats are often used in meditation, relaxation, and cognitive enhancement practices, as some believe they can influence brainwave patterns and mental states.

Bloch sphere The Bloch sphere is a geometric representation of the state of a two-level quantum system, such as a qubit. It is a sphere where any point on the surface represents a pure quantum state, and points inside the sphere represent mixed states.

Book of Mormon The Book of Mormon holds a special place in the world of religious texts due to its unique origins, distinctive theology, and significant impact on its followers. It is considered a sacred text by members of The Church of Jesus Christ of Latter-day Saints and other Latter-day Saint movement denominations. Believed to have been translated by Joseph Smith from golden plates in the early nineteenth century, the Book of Mormon offers a narrative of ancient American civilizations and their interactions with God, paralleling and complementing the Bible. Its role in shaping the faith, culture, and identity of millions of adherents underscores its profound influence and special status within religious literature.

Brainwaves Brainwaves are the electrical impulses in the brain that represent neural activity. They are generated by the synchronous activity of neurons and can be measured using an electroencephalogram (EEG). Brainwaves are categorized based on their frequency (speed) and amplitude (strength).

Brainwave patterns Brainwave patterns are the characteristic sequences and rhythms of brainwave activity that can be observed over time. These patterns can reflect various mental states, such as being awake and alert or being in deep sleep. Patterns can also indicate specific conditions or responses to stimuli.

Brainwave states Brainwave states refer to the different states of consciousness that are associated with various brainwave frequencies, such as delta (0.5–4 Hz) is associated with deep sleep and restorative states, and theta (4–8 Hz) is linked to light sleep, relaxation, and creativity, often present during daydreaming or meditation.

Cognitive faculties Cognitive faculties refer to the mental abilities or functions that enable individuals to process and understand information, think, reason, and engage in various intellectual tasks. These faculties encompass a wide range of cognitive processes and skills that are essential for human thinking and problem-solving.

Cognitive-behavioral therapy Cognitive-behavioral therapy (CBT) is a widely used, evidence-based form of psychotherapy that focuses on identifying and changing negative thought patterns and behaviors. Through structured sessions, CBT empowers individuals to take an active role in their recovery and build coping skills for long-term well-being. CBT aims to help individuals recognize distorted thinking, challenge irrational beliefs, and develop healthier ways of thinking and behaving.

Collective consciousness Collective consciousness refers to the set of shared beliefs, ideas, attitudes, and knowledge that are common to a group or a society. This concept, introduced by sociologist Émile Durkheim, suggests that collective consciousness shapes and unifies social groups, guiding their norms, values, and behaviors.

Complementary and alternative medicine The complementary and alternative medicine is a diverse range of medical practices, therapies, and products that are not typically part of conventional medicine. Complementary medicine is used with standard treatments to enhance their effectiveness, including practices like acupuncture and yoga. Alternative medicine is used in place of conventional treatments and includes approaches like herbal remedies and homeopathy. CAM aims to provide holistic care, focusing on physical, mental, and spiritual well-being.

Cortical thickness Cortical thickness is the distance between the outer surface (pial surface) and the inner surface (white matter boundary) of the brain's cerebral cortex. It is an important measure in neuroimaging studies, reflecting the density and health of neuronal layers in the cortex. Variations in cortical thickness are associated with developmental, aging, neurological, and psychiatric conditions, providing insights into brain structure and function.

Cross-frequency coupling analysis Cross-frequency coupling analysis examines the interaction between oscillatory signals at different frequencies. It assesses how the amplitude or phase of one frequency band modulates or influences the activity in another frequency band. This analysis is commonly used in neuroscience to explore how different brain wave frequencies interact, revealing insights into neural communication, cognitive processes, and the organization of brain networks.

Dream incubation Dream incubation is a technique used to influence the content of dreams. It involves focusing on a specific idea, question, or problem before falling asleep with the intention of dreaming about it. The process can include various practices such as writing down the desired dream topic, visualizing the scenario, repeating affirmations, or engaging in related activities before bed. Dream incubation aims to harness the subconscious mind's power to provide insights, solutions, or creative ideas through dreams.

DSM-IV criteria The DSM-IV, or the Diagnostic and Statistical Manual of Mental Disorders, Fourth Edition, is a classification system published by the American Psychiatric Association used by clinicians and researchers to diagnose and classify mental disorders. The DSM-IV provides standardized criteria for the diagnosis of mental health conditions, ensuring consistency and reliability in the identification and treatment of psychiatric disorders.

Ethereal energy Ethereal energy is a concept found in various spiritual and metaphysical traditions referring to a subtle, nonphysical form of energy that is believed to pervade the universe and influence all aspects of existence. This energy is often thought to be connected to the spiritual or metaphysical dimensions of reality, bridging the material world with the intangible realms.

False Discovery Rate Correction False Discovery Rate Correction is a statistical method used to control the expected proportion of false positives (incorrectly rejected null hypotheses) among the set of rejected hypotheses.

Gas Discharge Visualization The GDV (Gas Discharge Visualization) camera is a device developed by Dr. Konstantin Korotkov for capturing and analyzing the energy fields or "auras" around objects and people. It uses a technique where a high-voltage electrical field is applied to a subject, causing ionized gas emissions that are then captured on a special camera. This process creates images that are believed to reflect the subject's energy field or aura. The GDV camera can record, process, and interpret these images using computer software, providing detailed insights into the aura and its changes.

Generative AI models Generative AI models are sophisticated algorithms designed to produce new, original content by learning patterns from existing data. These models, such as GPT-4 and DALL-E, can generate text, images, music, and other types of media, often indistinguishable from human-created content. They function by using deep learning techniques, particularly neural networks, to understand and mimic the underlying structure of the data they are trained on. Generative AI models have a wide range of applications, from creating realistic art and enhancing creative processes to advancing natural language understanding and enabling more interactive and personalized technology experiences.

Graph theory approaches Graph theory approaches use mathematical structures known as graphs to model and analyze relationships and interactions within a network. A graph consists of nodes (or vertices) and edges (or links) that connect pairs of nodes.

Gray matter volume Gray matter volume refers to the amount of gray matter, which consists primarily of neuronal cell bodies, dendrites, and unmyelinated axons, in a specific region of the brain. It is often measured using neuroimaging techniques and is associated with various cognitive and motor functions.

Hebbian learning Hebbian learning is a neuropsychological theory that posits that neural connections are strengthened when the neurons on either side of the connection are activated simultaneously. According to Hebbian learning, the repeated and persistent stimulation of one neuron by another enhances the synaptic efficacy between them, effectively reinforcing the connection.

Intelligent Assistive Technologies Intelligent Assistive Technologies are advanced tools and systems designed to help individuals with disabilities or impairments perform daily activities and improve their quality of life. These technologies leverage AI to adapt to users' needs and provide personalized support.

Large language models Large language models (LLMs) are advanced artificial intelligence models designed to understand and generate human-like text based on vast amounts of data. They are trained using deep learning techniques on diverse and extensive text corpora, enabling them to predict and generate coherent sentences, paragraphs, or even entire articles.

Latent Dirichlet Allocation Latent Dirichlet Allocation (LDA) is a generative statistical model that is used to classify text in a document to a particular topic. It is one of the most popular methods for topic modeling, a technique used to uncover the hidden thematic structure in a large corpus of text.

Lucid dreaming Lucid dreaming is a phenomenon where the dreamer becomes aware that they are dreaming while the dream is still ongoing. This awareness allows the dreamer to potentially exert some degree of control over the dream's characters, narrative, and environment. Lucid dreaming occurs during REM (Rapid Eye Movement) sleep, a phase of sleep characterized by rapid eye movements, increased brain activity, and vivid dreams. Lucid dreaming can be used for various purposes, such as problem-solving, rehearsing real-life scenarios, exploring creativity, or simply enjoying fantastical experiences.

Multivariate Pattern Analysis Multivariate Pattern Analysis (MVPA) is a statistical technique used to analyze complex datasets by examining patterns of activity across multiple variables or features simultaneously. MVPA is used to decode and interpret patterns of brain activity across various regions or conditions, enabling researchers to identify specific neural representations, cognitive states, or behavioral outcomes.

Neural network Neural network is a foundational concept in artificial intelligence and machine learning, inspired by the structure and function of the human brain.

They are composed of layers of interconnected nodes, or "neurons," which simulate the neural connections found in biological brains. Each neuron in a network processes input data by performing simple computations and passing the results to subsequent neurons.

Oscillometric blood pressure measurement Oscillometric blood pressure measurement is a noninvasive method used to determine blood pressure by detecting the oscillations in the arterial wall caused by the pulsatile flow of blood. It uses automated cuff that inflates to occlude the artery and then gradually deflates while sensors detect the pressure fluctuations in the cuff.

Permutation testing Permutation testing is a nonparametric statistical method used to assess the significance of an observed effect or difference by comparing it to a distribution of effects generated under the null hypothesis. It involves repeatedly shuffling or permuting the data labels or values to create a distribution of test statistics that would occur by chance.

Phase synchrony analysis Phase synchrony analysis assesses the degree to which oscillatory signals or brain waves are in phase with each other over time. It measures how well the timing of one signal's oscillations aligns with the other one, indicating the level of coherence or synchronization between them.

Pleasure-Arousal-Dominance The Pleasure-Arousal-Dominance (PAD) model is a psychological framework that describes emotions based on three core dimensions: Pleasure (the degree of pleasantness or unpleasantness), Arousal (the level of activation or excitement), and Dominance (the feeling of control or dominance).

Placebo effect The placebo effect refers to the phenomenon where patients experience real or perceived improvements in their health condition after receiving a treatment that has no therapeutic effect. This treatment is often a placebo—a substance with no active medical ingredients, such as a sugar pill or saline injection.

Prana Prana is a term from Hindu philosophy and Indian spiritual traditions referring to the vital life force or energy that sustains life. It is believed to be the fundamental energy that permeates the universe and is present in all living beings. Prana flows through the body via pathways called nadis and is vital for physical health, mental clarity, and spiritual growth. Practices like yoga and pranayama focus on harnessing and balancing prana to enhance well-being and spiritual development.

Preimplantation genetic testing Preimplantation genetic testing is a medical procedure used in conjunction with in vitro fertilization to screen embryos for genetic abnormalities before they are implanted into the uterus. It aims to ensure that only healthy embryos are selected for implantation, reducing the risk of genetic disorders and increasing the likelihood of a successful pregnancy.

Qi (or Chi) Qi (or Chi) is a concept from Eastern philosophies, especially in Chinese culture, referring to a vital life force or energy that flows through all living things. It is believed to be the fundamental energy that sustains life, influences health, and connects the body, mind, and spirit. Qi is central to practices like acupuncture, Tai Chi, and Qigong, where balancing and enhancing this energy is thought to improve well-being and harmony.

Quantum connection A quantum connection or energy link refers to the idea that particles or entities are interconnected at a fundamental level in ways that transcend classical physical interactions. This concept often draws from phenomena observed in quantum mechanics, such as entanglement, where particles become linked and the state of one particle instantly influences the state of another, regardless of the distance between them.

Quantum entanglement Particles can become entangled, meaning that the state of one particle instantly influences the state of another, regardless of the distance separating them.

Quantum gates Quantum gates are fundamental operations in quantum computing that manipulate the state of qubits, the basic units of quantum information. They are analogous to classical logic gates but operate according to the principles of quantum mechanics. Quantum gates perform transformations on qubits by changing their probabilities and phases, enabling quantum algorithms to process and manipulate information in ways that classical gates cannot.

Relational I-Thou "Relational I-Thou" is a concept from the philosophy of dialogue introduced by Martin Buber, a twentieth-century Jewish philosopher. It describes a deep, meaningful relationship between individuals, characterized by mutual presence, openness, and genuine connection.

Remote patient monitoring Remote patient monitoring (RPM) is the process of using digital technologies to collect health data from patients in one location and securely transmit it to healthcare providers in a different location for assessment and recommendations. This data contains vital signs like blood pressure, heart rate, blood glucose levels, and other health metrics. RPM enables continuous monitoring and management of chronic conditions, early detection of potential health issues, and timely intervention, which can lead to better patient outcomes and reduced hospital admissions.

Seasonal affective disorder Seasonal affective disorder (SAD) is a type of depression that occurs at specific times of the year, usually during the fall and winter months when there is less natural sunlight. Symptoms include low mood, lack of energy, changes in sleep patterns, and difficulty concentrating. SAD is believed to be related to changes in light exposure, which can affect serotonin levels and circadian rhythms.

Selective serotonin reuptake inhibitors Selective serotonin reuptake inhibitors (SSRIs) are a class of medications commonly used to treat depression and anxiety disorders. They work by blocking the reabsorption (reuptake) of serotonin into neurons, making more serotonin available in the brain. This increase in serotonin levels helps improve mood, reduce anxiety, and enhance overall mental well-being. SSRIs are often prescribed because they tend to have fewer side effects compared to other antidepressants.

Singing bowls Singing bowls are metal or crystal bowls that produce a resonant, harmonious sound when struck or rubbed with a mallet. They are often used in meditation, sound therapy, and spiritual practices. The sound from singing bowls

Glossary

creates a soothing, vibrating tone that is believed to promote relaxation, balance, and healing by resonating with the body's energy centers. Each bowl can produce various tones and frequencies, depending on its size, material, and craftsmanship.

Sudarshan Kriya Sudarshan Kriya is a powerful rhythmic breathing technique developed by Sri Sri Ravi Shankar, the founder of the Art of Living Foundation. It is designed to harmonize the body, mind, and spirit, promoting overall well-being.

Symbolic AI Symbolic AI, also known as classical AI or good old-fashioned AI (GOFAI), is an approach to artificial intelligence that emphasizes the use of human-readable symbols to represent problems and logic to solve them. This method relies on the manipulation of symbols and the explicit encoding of all knowledge in a form that computers can process.

Synesthesia Synesthesia is a neurological condition where stimulation of one sensory pathway leads to automatic, involuntary experiences in another sensory pathway. For example, a person with synesthesia might see colors when they hear music or associate specific tastes with certain words. This blending of the senses results in a unique perceptual experience, where multiple senses are simultaneously activated.

TeleHealth TeleHealth refers to the use of digital communication technologies to deliver healthcare services and information remotely. This includes video consultations, remote monitoring, electronic health records, and mobile health apps. TeleHealth enables patients to access medical care from the comfort of their homes, improving convenience and accessibility, particularly for those in remote or underserved areas.

TF-IDF TF-IDF, which stands for Term Frequency-Inverse Document Frequency, is a statistical measure used to evaluate the importance of a word in a document relative to a collection of documents (corpus). It is commonly used in information retrieval and text mining to help identify which words are most significant in a document.

Vegetative state A vegetative state is a medical condition in which a person is awake but not aware of their surroundings. Individuals in this state may have sleep-wake cycles and exhibit basic reflexes, but they lack conscious awareness and purposeful behavior. This condition often results from severe brain injury, and while the person may open their eyes, respond to stimuli with reflex movements, and breathe without assistance, they do not show signs of cognitive function or conscious thought.

Zero-shot learning Zero-shot learning is a method in machine learning where a system learns to recognize objects, understand concepts, or perform tasks that it has not explicitly seen during training. This is fundamentally different from traditional machine learning techniques where the system learns from a large dataset of labeled examples for each category it needs to recognize or task it needs to perform.

References

Acabchuk, R. L., Simon, M. A., Low, S., Brisson, J. M., & Johnson, B. T. (2021). Measuring meditation progress with a consumer-grade EEG device: Caution from a randomized controlled trial. *Mindfulness, 12*, 68–81.

Afonso, R. F., Kraft, I., Aratanha, M. A., & Kozasa, E. H. (2020). Neural correlates of meditation: a review of structural and functional MRI studies. *Frontiers in Bioscience-Scholar, 12*(1), 92–115.

Ahmadi, M., Nopour, R., & Nasiri, S. (2023). Developing a prediction model for successful aging among the elderly using machine learning algorithms. *Digital Health, 9*, 1–22. Cited by: 2; All Open Access, Gold Open Access, Green Open Access.

Akdevelioglu, D., Hansen, S., & Venkatesh, A. (2022). Wearable technologies, brand community and the growth of a transhumanist vision. In *Transhumanisms and Biotechnologies in Consumer Society* (pp. 171–206). Routledge.

Alkhouri, K. I. (2024). The role of artificial intelligence in the study of the psychology of religion. *Religions, 15*(3), 1–27. Cited by: 0; All Open Access, Gold Open Access.

Alpaslan, C. M., & Mitroff, I. I. (2024a). Moral foundations of spirituality and religion through natural language processing. *Journal of Management, Spirituality & Religion, 21*(2), 184–205.

Alpaslan, C. M., & Mitroff, I. I. (2024b). Spiritual versus religious: A natural language processing perspective. *Journal of Management, Spirituality & Religion, 21*(1), 63–82.

Altammami, S., & Atwell, E. (2022). Challenging the transformer-based models with a classical Arabic dataset: Quran and hadith. In *Proceedings of the Thirteenth Language Resources and Evaluation Conference* (pp. 1462–1471). European Language Resources Association.

Amram, Y. (2007). The seven dimensions of spiritual intelligence: An ecumenical, grounded theory. In *115th Annual Conference of the American Psychological Association, San Francisco, CA* (Vol. 12)

Baars, B. J. (2005). Global workspace theory of consciousness: Toward a cognitive neuroscience of human experience. *Progress in Brain Research, 150*, 45–53.

Badran, B. W., Austelle, C. W., Smith, N. R., Glusman, C. E., Froeliger, B., Garland, E. L., Borckardt, J. J., George, M. S., & Short, B. (2017). A double-blind study exploring the use of transcranial direct current stimulation (tDCS) to potentially enhance mindfulness meditation (e-meditation). *Brain Stimulation: Basic, Translational, and Clinical Research in Neuromodulation, 10*(1), 152–154.

Baer, R., Gu, J., & Strauss, C. (2022). Five facet mindfulness questionnaire (FFMQ). In *Handbook of assessment in mindfulness research* (pp. 1–23). Springer.

Beck, U. (2011). Cosmopolitanism as imagined communities of global risk. *American Behavioral Scientist, 55*(10), 1346–1361.

Berne, S. (2012). The science of measuring wellness. *Journal of Behavioral Optometry, 23*(3), 1–27.

Bernstein, J. (1985). Out of my mind: Quantum reality. *The American Scholar, 54*(1), 7–14.

Betancourt, M. (2006). The aura of the digital. *CTheory. net, 9*.

Bohm, D. (2005). *Wholeness and the implicate order*. Routledge.

Bond, R. R., Mulvenna, M. D., Potts, C., O'Neill, S., Ennis, E., & Torous, J. (2023). Digital transformation of mental health services. *npj Mental Health Research, 2*(1), 13.

Brown, K. W. & Ryan, R. M. (2003). Mindful attention awareness scale. *Journal of Personality and Social Psychology*, 240.

Byrne, R. (2008). *The secret*. Simon and Schuster.

Callister, P., & Dykeman, C. (2021). The topic structure of the Book of Mormon: Worldview insights for mental health professionals. Oregon State University.

Capra, F. (2010). *The Tao of physics: An exploration of the parallels between modern physics and eastern mysticism*. Shambhala Publications.

Chalmers, D. (2017). The hard problem of consciousness. *The Blackwell Companion to Consciousness, 32*–42.

Chandra, R., & Ranjan, M. (2022). Artificial intelligence for topic modelling in Hindu philosophy: Mapping themes between the Upanishads and the Bhagavad Gita. *Plos One, 17*(9), e0273476.

Chandra, R., Tiwari, A., Jain, N., & Badhe, S. (2024). Large language models for metaphor detection: Bhagavad Gita and sermon on the mount. *IEEE Access, 12*, 84452.

Chen, Q., Lu, Y., Gong, Y., & Xiong, J. (2023). Can AI chatbots help retain customers? Impact of AI service quality on customer loyalty. *Internet Research, 33*(6), 2205–2243.

Clayton, P. (2004). *Mind and emergence: From quantum to consciousness*. OUP Oxford.

Connelly, K., Molchan, H., Bidanta, R., Siddh, S., Lowens, B., Caine, K., Demiris, G., Siek, K., & Reeder, B. (2021). Evaluation framework for selecting wearable activity monitors for research. *Mhealth, 7*, 1–13.

Cushman, S. (2023). John Muir's Thoreau. *Literary Imagination, 25*(2), 85–105.

Davis, J. J. J., Kozma, R., & Schübeler, F. (2023). Analysis of meditation vs. sensory engaged brain states using Shannon entropy and Pearson's first skewness coefficient extracted from EEG data. *Sensors, 23*(3), 1–23. Cited by: 3; All Open Access, Gold Open Access, Green Open Access.

Debong, F., Mayer, H., & Kober, J. (2019). Real-world assessments of mySugr mobile health app. *Diabetes Technology & Therapeutics, 21*(S2), S2–35.

Deka, B., & Deka, D. (2023). Nonlinear analysis of heart rate variability signals in meditative state: A review and perspective. *BioMedical Engineering OnLine, 22*(1), 35.

Ding, M., Zhao, A., Lu, Z., Xiang, T., & Wen, J.-R. (2019). Face-focused cross-stream network for deception detection in videos. In *Proceedings of the IEEE/CVF Conference on Computer Vision and Pattern Recognition* (pp. 7802–7811).

Ding, Y. (2020). The way of the Christianity Sinicization from the view of bible translation. *Dialogo, 7*(1), 103–111.

Dispenza, J. (2014). *You are the placebo: Making your mind matter*. Hay House.

Divarco, R., Ramasawmy, P., Petzke, F., & Antal, A. (2023). Stimulated brains and meditative minds: A systematic review on combining low intensity transcranial electrical stimulation and meditation in humans. *International Journal of Clinical and Health Psychology, 23*(3), 100369.

Duschinsky, R. (2012). Tabula rasa and human nature. *Philosophy, 87*(4), 509–529.

Elbers, J., & McCraty, R. (2020). Heartmath approach to self-regulation and psychosocial well-being. *Journal of Psychology in Africa, 30*(1), 69–79.

Elrod, A. (2024). Uncovering theological and ethical biases in LLMs: An integrated hermeneutical approach employing texts from the Hebrew Bible. *HIPHIL Novum, 9*(1):2–45.

Emmons, R. A. (2003). *The psychology of ultimate concerns: Motivation and spirituality in personality*. Guilford Press.

Farooq, M. S., & Salam, M. (2021). Cleaner production practices at company level enhance the desire of employees to have a significant positive impact on society through work. *Journal of Cleaner Production, 283*, 1–36. Cited by: 6; All Open Access, Green Open Access.

Feinberg, T. E. (2012). Neuroontology, neurobiological naturalism, and consciousness: A challenge to scientific reduction and a solution. *Physics of Life Reviews, 9*(1), 13–34.

Flanagan, K., & Saikia, M. J. (2023). Consumer-grade electroencephalogram and functional near-infrared spectroscopy neurofeedback technologies for mental health and wellbeing. *Sensors, 23*(20), 8482.

Ford, M. (2018). *Architects of Intelligence: The truth about AI from the people building it*. Packt Publishing Ltd.

Freud, S. (1989). *On dreams*. WW Norton & Company.

Fuller, R. C. (2001). *Spiritual, but not religious: Understanding unchurched America*. Oxford University Press.

Gambhirananda, S. (1980). *Katha Upanishad: With the Commentary of Shankaracharya*. Advaita Ashrama (A Publication House of Ramakrishna Math, Belur Math).

Gardner, H. (2000). A case against spiritual intelligence. *The International Journal for the Psychology of Religion, 10*(1), 27–34.

Garg, M. (2023). Mental health analysis in social media posts: A survey. *Archives of Computational Methods in Engineering, 30*(3), 1819–1842.

Garg, M. (2024). Mental disturbance impacting wellness dimensions: Resources and open research directions. *Asian Journal of Psychiatry, 92*, 103876.

Garg, M., Kumar, D., Samanta, D. S., & Sathiyaseelan, A. (2024a). *Impact of climate change on social and mental well-being*. Elsevier.

Garg, M., Sathvik, M., Raza, S., Chadha, A., & Sohn, S. (2024b). Reliability analysis of psychological concept extraction and classification in user-penned text. In *Proceedings of the International AAAI Conference on Web and Social Media* (Vol. 18, pp. 422–434).

Garg, M., Saxena, C., Krishnan, V., Joshi, R., Saha, S., Mago, V., & Dorr, B. J. (2022). CAMS: An annotated corpus for causal analysis of mental health issues in social media posts. In *Proceedings of Language Resources and Evaluation*.

Garg, M., Saxena, C., Samanta, D., & Dorr, B. J. (2023a). LonXplain: Lonesomeness as a consequence of mental disturbance in Reddit posts. In *International Conference on Applications of Natural Language to Information Systems* (pp. 379–390). Springer.

Garg, M., Shahbandegan, A., Chadha, A., & Mago, V. (2023b). An annotated dataset for explainable interpersonal risk factors of mental disturbance in social media posts. *Findings of the Association for Computational Linguistics (ACL). Conference Paper*, 11960.

Gazarian, P. K., Cronin, J., Dalto, J. L., Baker, K. M., Friel, B. J., Bruce-Baiden, W., & Rodriguez, L. Y. (2019). A systematic evaluation of advance care planning patient educational resources. *Geriatric Nursing, 40*(2), 174–180.

Gerson, L. P. (1986). Platonic dualism. *The Monist, 69*(3), 352–369.

Giannone, D. A., & Kaplin, D. (2020). How does spiritual intelligence relate to mental health in a western sample? *Journal of Humanistic Psychology, 60*(3), 400–417.

Goleman, D. (1996). Emotional intelligence. Why it can matter more than IQ. *Learning, 24*(6), 49–50.

Goleman, D. (2020). *Emotional intelligence*. Bloomsbury Publishing.

Grabb, D., Lamparth, M., & Vasan, N. (2024). Risks from language models for automated mental healthcare: Ethics and structure for implementation. *medRxiv*, 2024–04.

Gruber, M., Peltonen, J., Bartsch, J., & Barzyk, P. (2022). The validity and reliability of counter movement jump height measured with the polar vantage v2 sports watch. *Frontiers in Sports and Active Living, 4*, 1013360.

Hawkins, D. R. (2014). *Power vs. force*. Hay House, Inc.

Hawkins, D. R. (2015). *Transcending the levels of consciousness: The stairway to enlightenment*. Hay House, Inc.

Heisenberg, W., & Bohr, N. (1958). Copenhagen interpretation. *Physics and Philosophy, 16*, 39–53.

Hicks, E., & Hicks, J. (2009). *Ask and it is given: Learning to manifest your desires*. ReadHowYouWant.com.

Ho, J. K., & Hoorn, J. F. (2022). Quantum affective processes for multidimensional decision-making. *Scientific Reports, 12*(1), 20468.

Hunkin, H., King, D. L., & Zajac, I. T. (2021). EEG neurofeedback during focused attention meditation: Effects on state mindfulness and meditation experiences. *Mindfulness, 12*, 841–851.

Hutchinson, B. (2024). Modeling the sacred: Considerations when using religious texts in natural language processing. In *Findings of the Association for Computational Linguistics: NAACL 2024* (pp. 1029–1043).

Ingendoh, R. M., Posny, E. S., & Heine, A. (2023). Binaural beats to entrain the brain? A systematic review of the effects of binaural beat stimulation on brain oscillatory activity, and the implications for psychological research and intervention. *Plos One, 18*(5), e0286023.

Inkster, B., Sarda, S., Subramanian, V., et al. (2018). An empathy-driven, conversational artificial intelligence agent (Wysa) for digital mental well-being: real-world data evaluation mixed-methods study. *JMIR mHealth and uHealth, 6*(11), e12106.

Jacobs, G. (2009). Influence and canonical supremacy: An analysis of how George Herbert mead demoted Charles Horton Cooley in the sociological canon. *Journal of the History of the Behavioral Sciences, 45*(2), 117–144.

Jain, T., Lu, R. J., & Mehrotra, A. (2019). Prescriptions on demand: the growth of direct-to-consumer telemedicine companies. *JAMA, 322*(10), 925–926.

Janeczko, Z., & Foster, M. E. (2022). A study on human interactions with robots based on their appearance and behaviour. In *Proceedings of the 4th Conference on Conversational User Interfaces* (pp. 1–6).

Judith, A. (2011). *Eastern body, western mind: Psychology and the chakra system as a path to the self*. Celestial Arts.

Kamath, C. (2013). Analysis of heart rate variability signal during meditation using deterministic-chaotic quantifiers. *Journal of Medical Engineering & Technology, 37*(7), 436–448.

Kang, A., Le, T., & Chen, Y. (2024). Toshakhana: A multidimensional Panjabi corpus in Gurumukhi script. In *Proceedings of the 2024 ACM Southeast Conference* (pp. 278–283).

Kappattanavar, A. M., Hecker, P., Moontaha, S., Steckhan, N., & Arnrich, B. (2023). Food choices after cognitive load: An affective computing approach. *Sensors, 23*(14), 1–22. Cited by: 0; All Open Access, Gold Open Access.

King, D. B. (2008). Brighter paths to wellbeing: An integrative model of human intelligence and health. *Trent University Centre for Health Studies Showcase, 2008*, 12–13.

Kolirin, L. (2020). Talking robots could be used to combat loneliness and boost mental health in care homes. Available Online on CNN Health, 18 January 2022.

Kora, P., Meenakshi, K., Swaraja, K., Rajani, A., & Raju, M. S. (2021). EEG based interpretation of human brain activity during yoga and meditation using machine learning: A systematic review. *Complementary Therapies in Clinical Practice, 43*, 101329.

Krawczuk, D., Kulczyńska-Przybik, A., & Mroczko, B. (2024). Clinical application of blood biomarkers in neurodegenerative diseases—present and future perspectives. *International Journal of Molecular Sciences, 25*(15), 8132.

Krishna, A. B., et al., (2016). Spirituality and science of yogic chakra: A correlation. *Asian Journal of Complementary and Alternative Medicine, 4*(11), 17–22.

Kurzweil, R. (2000). *The age of spiritual machines: When computers exceed human intelligence*. Penguin.

Kwilecki, S. (2000). Spiritual intelligence as a theory of individual religion: A case application. *The International Journal for the Psychology of Religion, 10*(1), 35–46.

Laszlo, E. (2007). *Science and the Akashic field: An integral theory of everything*. Simon and Schuster.

Lazarus, R. S., & Lazarus, B. N. (1994). *Passion and reason: Making sense of our emotions*. Oxford University Press.

Lesage, M., Lavin, P., Rej, S., & Sekhon, H. (2023). Case series of a humanoid robot intervention for loneliness in long-term care homes. *The American Journal of Geriatric Psychiatry, 31*(3), S130–S131.

References

Liebeskind, C., Liebeskind, S., & Bouhnik, D. (2024). Machine translation for historical research: A case study of Aramaic-ancient Hebrew translations. *ACM Journal on Computing and Cultural Heritage, 17*(2), 1–23.

Lin, H., Han, Z., & Yang, L. (2024). Natural language processing-based comparative study of translating characteristics from the version of Tao Te Ching. *Advances in Education, Humanities and Social Science Research, 11*(1), 75–75.

Ling, X., Zhao, S., & Zhai, H. (2021). Quantum representation for robot's emotions based on pad model. In *The 7th International Workshop on Advanced Computational Intelligence and Intelligent Informatics*.

Manasa, B., Jois, S. N., & Prasad, K. N. (2020). Prana–the vital energy in different cultures: Review on knowledge and practice. *Journal of Natural Remedies, 20*(3), 128–139.

Mashhadimalek, M., Jafarnia Dabanloo, N., & Gharibzadeh, S. (2019). Is it possible to determine the level of spiritual well-being by measuring heart rate variability during the reading of heavenly books? *Applied Psychophysiology and Biofeedback, 44*, 185–193.

Mashour, G. A., & Alkire, M. T. (2013). Evolution of consciousness: Phylogeny, ontogeny, and emergence from general anesthesia. *Proceedings of the National Academy of Sciences, 110*(supplement_2), 10357–10364.

Maurya, M. K., Langeh, R., Anuradha, J., Sanjeevani, R., Sanjeevi, R., Tripathi, S., & Chauhan, D. (2021). Silica nanoparticles induced oxidative stress in different brain regions of male albino rats. *Scholars Academic Journal of Biosciences, 9*, 139–144.

May, A. D., & Maurin, E. (2021). Calm: A review of the mindful meditation app for use in clinical practice. *Families, Systems, & Health, 39*(2), 398–400.

McClafferty, H. (2017). *Integrative Pediatrics: Art, Science, and Clinical Application*. Cited by: 1.

Mercree, A. L. (2024). *Aura alchemy: Learn to sense energy fields, interpret the color spectrum, and manifest success*. Hay House, Inc.

Mills, A. (2009). Kirlian photography. *History of Photography, 33*(3), 278–287.

Miraglia, F. E. (2024). Unreliability of the gas discharge visualization (GDV) device and the bio-well for biofield science: Kirlian photography revisited and investigated. *Journal of Anomalistics Volume, 24*, 120–150.

Mostafa, M. M. (2018). Mining and mapping halal food consumers: Ageo-located twitter opinion polarity analysis. *Journal of Food Products Marketing, 24*(7), 858–879. Cited by: 47.

Munakata, Y., & Pfaffly, J. (2004). Hebbian learning and development. *Developmental Science, 7*(2), 141–148.

Myers, J. E., Sweeney, T. J., & Witmer, J. M. (2000). The wheel of wellness counseling for wellness: A holistic model for treatment planning. *Journal of Counseling & Development, 78*(3), 251–266.

Nair, V. K., Chandana Nair, B., Anupama, K., & Ajayan, C. (2024). Tech meets transcendence: ChatGPT and the next chapter of spiritual tourism. In *International Conference on Business and Technology* (pp. 100–114). Springer.

Ng, J. Y., Cramer, H., Lee, M. S., & Moher, D. (2024). Traditional, complementary, and integrative medicine and artificial intelligence: Novel opportunities in healthcare. *Integrative Medicine Research, 13*(1), 101024.

Nijakowski, K., Owecki, W., Jankowski, J., & Surdacka, A. (2024). Salivary biomarkers for Alzheimer's disease: A systematic review with meta-analysis. *International Journal of Molecular Sciences, 25*(2), 1168.

O'Daffer, A., Colt, S. F., Wasil, A. R., Lau, N., et al. (2022). Efficacy and conflicts of interest in randomized controlled trials evaluating headspace and calm apps: systematic review. *JMIR Mental Health, 9*(9), e40924.

Oman, D., & Morello-Frosch, R. (2018). Environmental health sciences, religion, and spirituality. In *Why religion and spirituality matter for public health: Evidence, implications, and resources* (pp. 139–152).

Oravec, J. A. (2022). The emergence of "truth machines"?: Artificial intelligence approaches to lie detection. *Ethics and Information Technology, 24*(1), 6.

OShea, R., Jones, M., & Getsoian, S. (2024). Measuring activity levels across the day using Philips health band watches: A case series. *Archives of Physical Medicine and Rehabilitation, 105*(4), e169.

Owayjan, M., Kashour, A., Al Haddad, N., Fadel, M., & Al Souki, G. (2012). The design and development of a lie detection system using facial micro-expressions. In *2012 2nd International Conference on Advances in Computational Tools for Engineering Applications (ACTEA)* (pp. 33–38). IEEE.

Palliya Guruge, C., Oviatt, S., Delir Haghighi, P., & Pritchard, E. (2021). Advances in multimodal behavioral analytics for early dementia diagnosis: A review. In *Proceedings of the 2021 International Conference on Multimodal Interaction* (pp. 328–340).

Papadopoulos, C., Hill, T., Battistuzzi, L., Castro, N., Nigath, A., Randhawa, G., Merton, L., Kanoria, S., Kamide, H., Chong, N.-Y., et al. (2020). The caresses study protocol: Testing and evaluating culturally competent socially assistive robots among older adults residing in long term care homes through a controlled experimental trial. *Archives of Public Health, 78*, 1–10.

Pashevich, E. (2023). Conceptualizing empathic child–robot communication. *Human–Machine Communication, 466*.

Piron, H. (2022). Meditation depth questionnaire (MEDEQ) and meditation depth index (MEDI). In *Handbook of Assessment in Mindfulness Research* (pp. 1–16). Springer.

Plante, T. G., & Sharma, N. K. (2001). Religious faith and mental health outcomes. *Faith and Health: Psychological Perspectives, 240*–261.

Possati, L. M. (2023). Psychoanalyzing artificial intelligence: The case of Replika. *AI & SOCIETY, 38*(4), 1725–1738.

Pradeep, A., Mamidi, R., & Satuluri, P. (2024). Context and WSD: Analysing Google Translate's Sanskrit to English output of Bhagavadgītā verses for word meaning. In *Proceedings of the 7th International Sanskrit Computational Linguistics Symposium* (pp. 14–26).

Prochaska, J. J., Vogel, E. A., Chieng, A., Kendra, M., Baiocchi, M., Pajarito, S., & Robinson, A. (2021). A therapeutic relational agent for reducing problematic substance use (Woebot): Development and usability study. *Journal of Medical Internet Research, 23*(3), e24850.

Puchalski, C. M. (2001). The role of spirituality in health care. In *Baylor University Medical Center Proceedings* (Vol. 14, pp. 352–357). Taylor & Francis.

Purdy, M., & Dupey, P. (2005). Holistic flow model of spiritual wellness. *Counseling and Values, 49*(2), 95–106.

Qahl, S. H. M. (2014). *An automatic similarity detection engine between sacred texts using text mining and similarity measures*. Rochester Institute of Technology.

Radin, D. (2009). *Entangled minds: Extrasensory experiences in a quantum reality*. Simon and Schuster.

Raghuvanshi, A., & Perkowski, M. (2010). Fuzzy quantum circuits to model emotional behaviors of humanoid robots. In *IEEE Congress on Evolutionary Computation* (pp. 1–8). IEEE.

Rubik, B. (2004). Scientific analysis of the human aura. In *Measuring energy fields state of the science*. Fair Lawn, NJ, Backbone (pp. 157–170).

Rudy, K. M. (2016). *Piety in pieces: How medieval readers customized their manuscripts*. Cited by: 33; All Open Access, Hybrid Gold Open Access.

Saghiri, A. M., Vahidipour, S. M., Jabbarpour, M. R., Sookhak, M., & Forestiero, A. (2022). A survey of artificial intelligence challenges: Analyzing the definitions, relationships, and evolutions. *Applied Sciences (Switzerland), 12*(8), 1–21. Cited by: 20; All Open Access, Gold Open Access.

Sah, P., & Fokoué, E. (2019). What do Asian religions have in common? An unsupervised text analytics exploration. Preprint. arXiv:1912.10847.

Sahel, S. S. D., & Boudour, M. (2019). Wavelet energy moment and neural networks based particle swarm optimisation for transmission line protection. *Bulletin of Electrical Engineering and Informatics, 8*(1), 10–20. Cited by: 7; All Open Access, Bronze Open Access, Green Open Access.

Sarithadevi, S., & Rajesh, R. (2023). *Character recognition for Malayalam palm leaf manuscripts: An overview of techniques and challenges* (Vol. 2773). Cited by: 0.

References

Schaffer, J. (2010). Monism: The priority of the whole. *Philosophical Review, 119*(1), 31–76.

Seetharaman, R., Avhad, S., & Rane, J. (2024). Exploring the healing power of singing bowls: An overview of key findings and potential benefits. *Explore, 20*(1), 39–43.

Sethi, N., Dev, A., Bansal, P., Sharma, D. K., & Gupta, D. (2023). Enhancing low-resource Sanskrit-Hindi translation through deep learning with Ayurvedic text. *ACM Transactions on Asian and Low-Resource Language Information Processing*.

Sharma, A. (2000). *Classical Hindu thought: An introduction*. Oxford University Press.

Shibata, T., Hung, L., Petersen, S., Darling, K., Inoue, K., Martyn, K., Hori, Y., Lane, G., Park, D., Mizoguchi, R., et al. (2021). Paro as a biofeedback medical device for mental health in the covid-19 era. *Sustainability, 13*(20), 11502.

Showail, A. J. (2022). Solving Hajj and Umrah challenges using information and communication technology: A survey. *IEEE Access, 10*, 75404–75427. Cited by: 4; All Open Access, Gold Open Access.

Shukla, A., Tiwari, S., & Hoang, V. T. (2023). Yoga practitioners and non-yoga practitioners to deal neurodegenerative disease in neuro regions. In *Data analysis for neurodegenerative disorders* (pp. 67–91). Springer.

Smith, J., & Skousen, R. (2022). *The book of Mormon: The earliest text*. Yale University Press.

Stuart, H. (1998). Quantum computation in brain microtubules? The Penrose–Hameroff 'Orch OR' model of consciousness. *Philosophical Transactions of the Royal Society of London. Series A: Mathematical, Physical and Engineering Sciences, 356*(1743), 1869–1896.

Sui, C. K. (2015). The ancient science and art of Pranic Healing. *(No Title)*.

Sweeney, T. J. (1998). *Adlerian counseling: A practitioner's approach*. Taylor & Francis.

Terman, L. M., & Merrill, M. A. (1960). Stanford-Binet intelligence scale: Manual for the third revision, form LM.

The National Wellness Institute (NWI). (2020). Six dimensions of wellness.

Thoresen, C. E. (1999). Spirituality and health: Is there a relationship? *Journal of Health Psychology, 4*(3), 291–300.

Tracy, B. (2005). *Change your thinking, change your life: How to unlock your full potential for success and achievement*. John Wiley & Sons.

Trepczyński, M. (2023). Religion, theology, and philosophical skills of LLM–powered chatbots. *Disputatio Philosophica: International Journal on Philosophy and Religion, 25*(1), 19–36.

van den Broek-Altenburg, E., Gramling, R., Gothard, K., Kroesen, M., & Chorus, C. (2021). Using natural language processing to explore heterogeneity in moral terminology in palliative care consultations. *BMC Palliative Care, 20*, 1–11.

van der Schyff, E. L., Ridout, B., Amon, K. L., Forsyth, R., & Campbell, A. J. (2023). Providing self-led mental health support through an artificial intelligence–powered chat bot (Leora) to meet the demand of mental health care. *Journal of Medical Internet Research, 25*, e46448.

Van Puyvelde, M., Neyt, X., McGlone, F., & Pattyn, N. (2018). Voice stress analysis: A new framework for voice and effort in human performance. *Frontiers in Psychology, 9*, 1994.

Varon, E. J. (1936). Alfred Binet's concept of intelligence. *Psychological Review, 43*(1), 32.

Vygotsky, L., & Cole, M. (2018). Lev Vygotsky: Learning and social constructivism. In *Learning theories for early years practice* (pp. 68–73). SAGE Publications Inc.

Wangmo, T., Lipps, M., Kressig, R. W., & Ienca, M. (2019). Ethical concerns with the use of intelligent assistive technology: Findings from a qualitative study with professional stakeholders. *BMC Medical Ethics, 20*, 1–11.

Wechsler, D. (1945). A standardized memory scale for clinical use. *The Journal of Psychology, 19*(1), 87–95.

Wilson, E. (2003). *The spiritual history of ice: Romanticism, science and the imagination*. Springer.

Wolman, R. (2001). *Thinking with your soul: Spiritual intelligence and why it matters*. Richard N. Wolman, PhD.

World Health Organization. (2022). World mental health report: Transforming mental health for all.

Yan, F., Iliyasu, A. M., & Hirota, K. (2021a). Conceptual framework for quantum affective computing and its use in fusion of multi-robot emotions. *Electronics, 10*(2), 100.

Yan, F., Iliyasu, A. M., Liu, Z.-T., Salama, A. S., Dong, F., & Hirota, K. (2015). Bloch sphere-based representation for quantum emotion space. *Journal of Advanced Computational Intelligence and Intelligent Informatics, 19*(1), 134–142.

Yan, F., Yang, X., Li, N., Yu, X., & Zhai, H. (2021b). Emotion generation and transition of companion robots based on Plutchik's model and quantum circuit schemes. *Security and Communication Networks, 2021*(1), 6802521.

Yang, X., Li, Y., Li, Q., Liu, D., & Li, T. (2022). Temporal-spatial three-way granular computing for dynamic text sentiment classification. *Information Sciences, 596*, 551–566.

Zadeh, A. R. (2023). Artificial intelligence and modern information technologies applications in Islamic sciences: A survey. *International Journal on Perceptive and Cognitive Computing, 9*(2), 48–61.

Zhang, T., Schoene, A. M., & Ananiadou, S. (2021). Automatic identification of suicide notes with a transformer-based deep learning model. *Internet Interventions, 25*, 1–8. Cited by: 17; All Open Access, Gold Open Access, Green Open Access.

Zhao, H. J., & Liu, J. (2018). Finding answers from the word of god: Domain adaptation for neural networks in biblical question answering. In *2018 International Joint Conference on Neural Networks (IJCNN)* (pp. 1–8). IEEE.

Zohar, D. (1997). *Rewiring the corporate brain: Using the new science to rethink how we structure and lead organizations*. Berrett-Koehler Publishers.

Zohar, D. (2012). *Spiritual intelligence: The ultimate intelligence*. Bloomsbury publishing.

Zukav, G. (2012). *The dancing Wu Li masters: An overview of the new physics*.

Zukav, G., & March, R. H. (1979). *The dancing Wu Li masters: An overview of the new physics*.

Printed and bound by CPI Group (UK) Ltd, Croydon, CR0 4YY
25/11/2024
01794112-0002